Auto Repair & Maintenance

ALPHA BOOKS

Published by Penguin Random House LLC

Penguin Random House LLC, 375 Hudson Street, New York, New York 10014, USA • Penguin Random House LLC (Canada), 90 Eglinton Avenue East, Suite 700, Toronto, Ontario M4P 2Y3, Canada (a division of Pearson Penguin Canada Inc.) • Penguin Books Ltd., 80 Strand, London WC2R 0RL, England • Penguin Ireland, 25 St. Stephen's Green, Dublin 2, Ireland (a division of Penguin Books Ltd.) • Penguin Random House LLC (Australia), 250 Camberwell Road, Camberwell, Victoria 3124, Australia (a division of Pearson Australia Group Pty. Ltd.) • Penguin Books India Pvt. Ltd., 11 Community Centre, Panchsheel Park, New Delhi—110 017, India • Penguin Random House LLC (NZ), 67 Apollo Drive, Rosedale, North Shore, Auckland 1311, New Zealand (a division of Pearson New Zealand Ltd.) • Penguin Books (South Africa) (Pty.) Ltd., 24 Sturdee Avenue, Rosebank, Johannesburg 2196, South Africa • Penguin Books Ltd., Registered Offices: 80 Strand, London WC2R 0RL, England

International Standard Book Number: 978-1-61564-762-0
Library of Congress Catalog Card Number: 2015930786

17 16 15 8 7 6 5 4 3 2 1

Interpretation of the printing code: The rightmost number of the first series of numbers is the year of the book's printing; the rightmost number of the second series of numbers is the number of the book's printing. For example, a printing code of 15-1 shows that the first printing occurred in 2015.

Printed in China

Publisher: Mike Sanders
Associate Publisher: Billy Fields
Senior Acquisitions Editor: Brook Farling
Development Editor: Ann Barton
Design Supervisor: William Thomas
Production Editor: Jan Lynn
Indexer: Johnna Dinse
Proofreader: Claudia Bell

Contents

CHAPTER 7
THE COOLING SYSTEM 121

CHAPTER 8
THE AIR CONDITIONING SYSTEM 141

CHAPTER 9
THE EXHAUST SYSTEM 155

CHAPTER 10
THE COMPUTER SYSTEM 167

CHAPTER 11
THE ELECTRICAL SYSTEM 177

CHAPTER 12
THE BRAKING SYSTEM 195

CHAPTER 13
STEERING AND SUSPENSION 221

CHAPTER 14

TIRES AND WHEELS 237

CHAPTER 15

THE EXTERIOR 255

APPENDIXES

INDEX 276

Introduction

"My car is doing something weird. What would cause that?"

That question, in a nutshell, is why I decided to write this book. As an owner of a business that deals with automobiles, I am constantly getting some form of that question from family, friends, and customers. Sometimes it is a difficult question that requires me to retreat to my library in search of answers, and sometimes it is common knowledge that everyone who owns a car should know.

When I began researching this book, I looked at what was already out there. So many books try to educate you to the point where you can describe exactly what is happening in a four-cycle engine. I think that is missing the point. You probably aren't looking for that much information; you just want to know why the car is making that funny rattle. Often the "expert" wants to share his or her entire knowledge base with you, and all you want to know is if you need to get that thing checked out. I don't typically respond to the above question with a full explanation of how the engine operates. Instead, I usually ask a follow-up question or two and then give you a few things to check. What you find will determine whether you can fix it yourself or if it is time to take it to a professional.

The other problem that arose in writing this book was the fact that automobiles are changing and becoming more sophisticated every year. What was common knowledge a few years ago is now ancient history, so keeping this book relevant was a challenge.

To this end, I decided to write this book as if I were answering your direct question about what is going on with your vehicle. I'll start by answering questions that will most likely pop up while you own your car. I've asked other experts about what they want you to know about your car when you bring it to them, and what you can do to prevent problems from arising. I won't cover everything that may happen, but I will touch on the most common problems. I'll give you some tips along the way on things you can do yourself, and how to stretch the life of an older car if you're on a budget.

The aim is that you'll keep this book to use with your future vehicles. If your friend who's good with cars isn't available, you'll have somewhere to go to for the answers you need. Or perhaps you're the uncle or parent who gives this book to a new driver who is in need of some basic knowledge about his or her car. Either way, I hope this book serves you as well as your automobile does.

ACKNOWLEDGMENTS

First and foremost, thanks goes to my mother, Carole Stribling, for editing my Midwestern mayhem. Thanks to my kids, Caleb Surber, Kayla Surber, Kara Stribling, and Jordan Stribling. To my sister, Marcia Broady, thanks for your belief in me. Thanks to the following individuals who also contributed information and/or their vehicles to write this book: Amy Stribling, Craig Pattee, Don Gould, Frank Frost, Jeff Etter, John Stephenson, Kenzie Warrick, Kim Stribling, Laurie Studebaker, Loraine Williams, Roger Walther, Marsha Sledge, Marti Henderson, Michael Melvin, Randal Starlin, Reed Oliver, Ron Dickerson, Ron Mendenhall, Seth Nichols, Stephen Maxwell, Steve Wright, Todd Fouty, and Wes Shelton.

CHAPTER 1

THE BASICS

IN THIS CHAPTER

What Every Driver Should Know

There's a lot of information in this book about how to take care of your car, and not all of it will apply to you. But as a car owner and driver, there are some things you need to know and some things you need to have with any vehicle you drive. Here are the things I, your mechanic, your emergency roadside service, your tire dealer, and your friend who is good with cars wants you to know when driving any vehicle. The rest is optional reading.

1. KNOW WHAT TO DO IN AN EMERGENCY (CHAPTER 2).
   You may have the best roadside service or automatic call-in on the planet, but it takes time for emergency and roadside help to arrive. You need to be able to secure yourself, your passengers, and your vehicle from harm.

2. UNDERSTAND HOW YOUR SAFETY SYSTEMS WORK.
   Car manufacturers are always improving the safety of their vehicles by adding features such as automatic braking and warning sounds to alert you if you drift out of your lane or if there is something in your blind spot. These features can be very helpful, but if you aren't aware of them or don't know how they work, they can cause panic or confusion when they engage. Familiarize yourself with your car's safety features so you know what to expect.

   It's also important to know how passenger restraint systems, like seat belts and airbags, work in your particular car. Height and distance restrictions are very important, especially for children.

3. CARRY YOUR OWNER'S MANUAL.
   The owner's manual contains a lot of information about your specific vehicle and what to do in emergency situations. Used cars may not always come with an owner's manual. If you don't have one, check with your dealer or go online and get one.

4. KNOW HOW TO INSPECT, MAINTAIN, AND CHANGE YOUR TIRES (CHAPTER 14).

   Carry a tire pressure gauge and inspect your tires regularly for signs of wear. Maintaining the correct air pressure will save fuel and prevent unnecessary wear. Regular inspections will alert you to any potential problems.

   Even if you never have to change your own tire, you should know how to do it correctly and have the necessary equipment in your car. You need to know where your spare is located and how to get it out. The location of the spare may make it challenging to remove, so check to see if you are capable of removing it and changing it yourself. If you buy a used car, make sure the spare is in good condition and be sure you have a jack and jack wrench in the car.

5. KNOW HOW TO CHECK THE FLUIDS IN YOUR CAR.

   Most people don't check their fluids as frequently as they should. Like checking your tires, checking your fluids can prevent problems down the road.

6. KNOW HOW TO CHECK AND REPLACE FUSES (CHAPTER 2).

   Consult your owner's manual so you know where to look if you have to change a fuse. Newer cars typically have two fuse boxes: one for high-power circuits and another for low-power circuits.

7. KNOW HOW TO CHECK COMPUTER CODES (CHAPTER 10).

   Your car can tell you more about itself than I can. If your car doesn't show you the codes on your computer display, know how to use a scanner and read the computer error codes. This way, you know if the repair guy is being honest with you, and whether or not the problem with your car is serious.

8. KNOW HOW TO PROPERLY JUMP-START YOUR VEHICLE (CHAPTER 2).

   As with changing a tire, you may never do this, but you need to know how to do it on your car. The battery isn't always in a convenient location, and it's easy to damage the electrical system if the car is jump-started incorrectly.

9. KNOW HOW TO FUEL YOUR CAR.

   There are a lot of fuels available today: gasoline, E-85, methanol blends, bio fuels, and high and low octane. Not all fuels are compatible with all vehicles. Check your owner's manual to see what kind of fuel your car is designed to use.

10. OWN A SHOP MANUAL.

    For those who own an older car, plan on keeping their car for a long time, or simply can't afford to take it to a professional, a shop manual is essential. Even if you never open it yourself, if you ask for help or advice from a friend who is good with cars, it is the first thing your helper will ask you to provide.

Types of Cars

Vehicles are categorized in two ways: how they are powered (by engine, battery, or a combination) and how that power is transferred to the wheels (the drivetrain). Taken together, the power source and the drivetrain are called the *power train*. It's important to know what kind of vehicle you have, because different vehicles perform and behave differently and require different maintenance.

POWER SOURCES

Most cars are powered in one of three ways: by a combustion engine, by one or more electric motors, or by a combination of both called a *hybrid*.

FUEL-BURNING ENGINES

The majority of vehicles on the road today use an engine to generate power by burning some type of fuel. The most common types of fuel are gasoline, diesel, hydrogen, and methanol. The advantage of fuel-burning engines is that they deliver plenty of power to move the car and generate lots of electricity to power computers, air conditioners, and other vehicle amenities. The disadvantage to fuel-burning engines is that most of them produce harmful emissions, and require additional equipment to help keep pollution to a minimum.

ELECTRIC MOTORS

Electric vehicles use energy stored in a battery or batteries, which power one or more electric motors to drive the wheels. Because electric motors don't burn fuel, they do not generate any harmful emissions. The downside to electric cars is that they have limited range, and using energy-intensive features like air conditioning and power windows can drain power quickly. Electric batteries are also expensive to replace.

HYBRIDS

A hybrid vehicle has a battery (or batteries) and an electric motor as well as a fuel-burning engine. The electric motor can be used to drive the wheels, and the engine can both generate electricity to charge the battery and supply power to drive the wheels. This dual power source results in greater engine efficiency and reduces emissions. Hybrids come in many sizes and rival conventional engine-driven cars in abilities.

ELECTRIC VEHICLE DRIVETRAINS

Electric vehicles can be front-wheel, rear-wheel, or four-wheel drive. Electric vehicles may or may not use a transmission, and may have more than one motor driving the wheels. You might have a single electric motor driving the rear wheels through a transmission or four separate motors driving each wheel directly without a transmission.

DRIVETRAINS

The drivetrain is the mechanism used to transfer the power generated by the engine or motor to the wheels that move the car. The components of the drivetrain vary depending on the orientation of the engine in relationship to the driving wheels. Most cars on the market today use one of three configurations: rear-wheel drive, front-wheel drive, or four-wheel drive (also called all-wheel drive).

REAR-WHEEL DRIVE

In a rear-wheel-drive vehicle, the rear wheels are the "drive" wheels. They receive the power from the engine through the drivetrain and "push" the car, while the front wheels are used for steering. Most rear-wheel-drive vehicles have the engine at the front of the car, which means the power to drive the wheels travels through a transmission, a drive shaft, and a set of gears before reaching the rear wheels. This layout makes rear-wheel-drive vehicles easier to maintain and build as there is more room to spread things out, and the front wheels are less complex. However, rear-wheel-drive vehicles are generally less fuel-efficient than front-wheel-drive cars.

FRONT-WHEEL DRIVE

In a front-wheel-drive vehicle, the front wheels are the "drive" wheels. They receive power from the engine, which is typically located at the front of the car. This allows for better weight distribution in the car, and the weight of the engine helps the traction of the wheels and reduces the amount of power lost through long drivetrains. However, front-wheel-drive cars can be mechanically complicated because the drive axles and the steering system are all together. Front-wheel drive is typically used on smaller, more fuel-efficient cars.

FOUR-WHEEL OR ALL-WHEEL DRIVE

In this configuration, the engine sends power to all four wheels of the vehicle. The engine is located at the front of the car in most all-wheel vehicles. When all four wheels receive power, the car can handle better in poor driving conditions. However, this type of drivetrain adds complexity to the car and reduces efficiency. Larger trucks that drive in off-road conditions and sport utility vehicles benefit from all wheel drive.

A Look Under the Hood

Every car is different, but this quick guide will give you a general idea of where things are located on front-, rear-, and four-wheel-drive vehicles, as well as hybrid vehicles. Check your owner's manual for the location of components in your specific vehicle.

FRONT-WHEEL DRIVE

A typical front-wheel-drive car has the engine on the left and the transaxle on the right.

1. Oil fill location
2. Oil dipstick (under cover)
3. Coolant reservoir
4. Brake master cylinder and brake fluid fill location
5. Transmission fluid dipstick
6. Power steering fluid reservoir
7. Air filter
8. Fuse box
9. Windshield washer fill location
10. Battery positive post
11. Drive belts (under cover)
12. Radiator fill cap

FRONT-WHEEL DRIVE

This V6 car does not have a transmission dipstick tube, and it uses electronic power steering, so there is no power steering reservoir.

1. Oil fill location
2. Oil dipstick
3. Coolant reservoir
4. Brake master cylinder and brake fluid fill location
5. Air filter
6. Windshield washer fill location
7. Battery positive post
8. Drive belts (under cover)
9. Radiator fill cap

FOUR-WHEEL DRIVE AND REAR-WHEEL DRIVE

This four-wheel-drive truck has the engine is located right in front. It has two batteries, one on each side of the radiator. The engine drive belts are on the front of the motor. Most rear-wheel-drive vehicles have a similar layout.

1. Oil fill location
2. Oil dipstick
3. Coolant reservoir
4. Brake master cylinder and brake fluid fill location
5. Transmission fluid dipstick
6. Power steering reservoir (below the hoses)
7. Air filter
8. Fuse box
9. Battery positive post (two on this vehicle)
10. Drive belts (under cover)
11. Radiator fill cap

HYBRID

Hybrids use a coolant system for the batteries that isn't found on typical engine-only cars. This hybrid does not use hydraulic power steering. Since the batteries are remotely located, a charge point is located in the fuse box.

1. Oil fill location
2. Oil dipstick
3. Coolant reservoir (under cover)
4. Brake master cylinder and brake fluid fill location
5. Transmission fluid dipstick
6. Air filter
7. Fuse box
8. Windshield washer fill location
9. Battery positive post (inside the fuse box)
10. Drive belts
11. Radiator fill cap
12. Battery coolant (hybrid only)

Identifying Your Car's Fluids

You've just found a puddle under your car. What is it? Is it something serious or even dangerous?

Your car uses fluids to do all kinds of things, and occasionally they leak out of the car. It's important to know what the different fluids look, feel, and smell like, and what to do if you find a puddle on your driveway.

OILS

Your car uses oils to lubricate metal pieces, allowing them to move against each other without friction. Oil may be made from petroleum or synthetically produced. The color of engine oil will differ depending on the manufacturer and its type. **Clean engine oil** is usually tan or light brown in color. **Heavy gear oil** is darker in color and thick like honey. **Used motor oil** is blackened by the deposits left in the engine after burning fuel. Oil is very slick when you touch it and has a smell of sulfur or burnt popcorn.

A car that is leaking oil will eventually fail when the metal parts are no longer able to move against each other smoothly. Oil leaks can be caused by engine or axle seals failing, a clog in the engine vent, a loose oil filter or drain plug, or even spilling oil on the engine while filling it up.

If you find oil, try to locate the source of the leak. If the leak is from the engine, check your oil level. If the level is okay and the leak is small, you can take the car in to have it checked out by a professional. If the leak is big or your engine is low on oil because of the leak, have the car towed to a mechanic—don't trust that it will make it there on its own.

BRAKE FLUID OR HYDRAULIC FLUID

Hydraulic fluid is used to operate the brakes and sometimes the power steering. Hydraulic fluid is designed to be compressed, creating high pressure to operate various devices. Brake fluid is clear or light amber in color, and is the consistency of water, but it is slick like oil. It smells a little like fish or baby oil.

If you find brake fluid under your car, do not drive it. Leaking brakes are dangerous, and if the hydraulic fluid is leaking from your power steering system, it can be destroyed very quickly. Have your car towed to a mechanic.

TRANSMISSION AND TRANSAXLE FLUIDS

Transmission and transaxle fluids work like both oils and hydraulic fluids. They lubricate the metal parts and can be pressurized like hydraulic fluids to operate the transmission while shifting. They are sometimes used in place of hydraulic fluids in power steering systems.

Transmission fluids are usually dyed to distinguish them from other fluids. Most often they are dyed red, but some manufacturers use other colors. The bright color is the easiest way to identify transmission fluids. They are oily but usually have a thinner consistency than oils, and they have a burnt smell.

If you find transmission fluid, locate the leak and determine if it was caused by overfilling. A very small leak can be driven to the mechanic. If it is a big leak, do not drive the car.

ANTIFREEZE

Antifreeze, or coolant, is mixed with water and is designed to remove the heat from your engine. At the same time, it prevents the water in your cooling system from freezing and expanding, which will tear apart metal and destroy the engine.

Antifreeze comes in a rainbow of colors, but most of them are a fluorescent color. Neon yellow and green are the most common. It has a very sweet smell to it, and it feels slick like soapy water.

An antifreeze leak means that a seal has failed or the engine is getting too hot, and the coolant is bubbling out of the system. Because the hot coolant expands, the system builds pressure, so be careful when checking a hot engine for leaks—coolant can scald. If the coolant is coming out of the overflow reservoir, your engine is probably running too hot, and you need to have it looked at by a mechanic. If it is a very slow leak, check for drips around hoses and fittings. If you have a high-pressure leak that streams out of an area due to corrosion or a seal failure, have your car towed to the mechanic.

WATER

Water can drip off of your car for a number of reasons, most of them harmless. Most often, dripping water is coming from condensation on the air conditioner. Water may also condense in the exhaust system and be blown out of the tail pipes before the exhaust is hot enough to evaporate it. Water can also collect in areas after washing or driving through wet conditions and be jostled out of place at a later time.

You need to be concerned if your engine cooling system is leaking straight water, which may mean that that the system does not have any antifreeze. Your engine cannot run on straight water.

If water is coming out of the transmission or any other system, you have a leak that is allowing water to get in. You need to have this looked at because water corrodes and can be destructive.

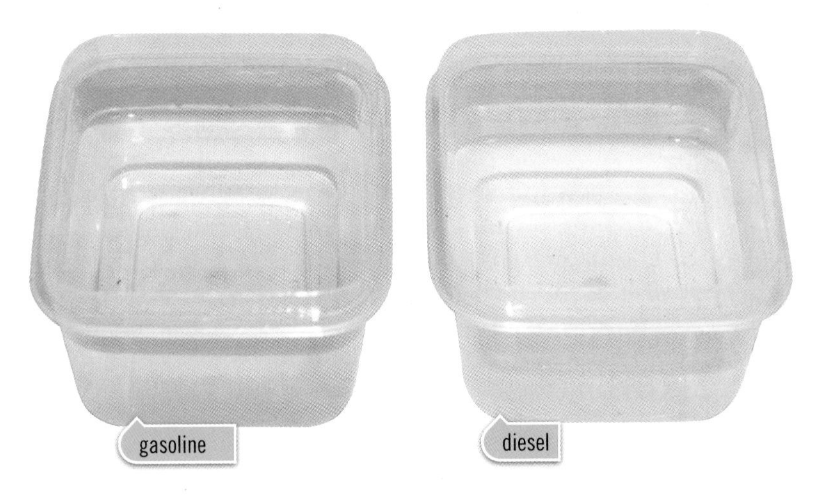

TIP

Your leak might be hard to identify if it's contaminated with road debris. If you find a puddle, place a clean piece of cardboard under the leak area to help locate the point of the leak. The cardboard may help you identify the type of fluid that is leaking before it hits the ground.

FUELS

The most common fuels today are gasoline and diesel. Fuels have the consistency of water and can be clear to amber in color, depending on the grade of the fuel and its refinement. Diesel fuel has a slight oily feel to it. Fuels are known for their strong odors, which is a good indicator of a fuel leak. Bio-fuels may smell like cooking oil or burnt popcorn depending on what they are made of.

If you find a fuel leak that is caused by something other than overfilling the tank, do NOT drive the car. Fuel leaks are very dangerous and potentially explosive. Call a mechanic if you find a fuel leak in your car.

WINDSHIELD WASHER FLUID

Windshield washer fluid is usually blue, but some manufacturers use other colors to differentiate their products. Most fluids are good for de-icing to a certain temperature, and some help to remove stubborn debris from your windshield. Washer fluid has the consistency of water and may have a slight cleanser odor or sweet smell to it.

Leaking washer fluid is nothing to panic about. Your car will keep running if it runs out of fluid, although it isn't a good idea to run the system dry for too long. You can have the leak fixed at your convenience.

Finding a Good Mechanic

Finding a reputable mechanic can be a source of anxiety for many people. If you're not knowledgeable about cars, taking your car in for service can be intimidating. You may not know what to expect or worry that you're not getting a fair deal. There's no foolproof way to be sure your mechanic or other car specialist will repair your car correctly, but there are some things to keep in mind as you consider your options.

GO TO THE DEALER IF YOUR CAR IS UNDER WARRANTY.

In general, if you have a car under warranty, your best option is to take it to the dealer for repairs. Having the car serviced by someone else may void your warranty, and it's in the best interest of the dealership to do a good job to encourage you to come back after the warranty expires.

LOOK FOR CERTIFICATIONS.

Some systems, like air conditioning, require a person to be certified before they are authorized to purchase parts and work on the system. Certifications are a good thing—they tell you that someone has been trained to work on your car properly—but they aren't necessarily a guarantee of quality work.

CHECK FOR AFFILIATIONS.

More and more insurance companies are partnering with repair shops and some are opening up their own. A shop that is affiliated with your auto club or roadside service is likely to have been vetted for reliable work.

ASK ABOUT PARTS.

Don't assume that a dealership will use original equipment parts for your repair. If you're concerned about the use of aftermarket parts, you can specify the use of original equipment. However, this may be more expensive.

Don't be afraid to ask to see the parts that were removed from your car after they've been replaced. In many cases the mechanic will be able to show you what failed.

CONSIDER CAPABILITY.

Dealers usually have the ability to work on new vehicles, but they may not be the best choice for working on older vehicles or other makes. If you have an older car with a carburetor, the young mechanics in the dealership may not know anything about it. Likewise, your uncle who fixes his own cars may not have the tools required to handle your new, computer-controlled car. Make sure the pros are equipped to do the job.

WHEN YOU TALK TO YOUR MECHANIC

- Be prepared to explain the problems you're experiencing in detail and tell the mechanic what the conditions were when the problem began.
- Research as much as you can before dealing with your mechanic. The more you know, the more easily you'll be able to discuss the problem.
- Don't hesitate to get estimates from a couple shops, but be courteous and respectful when discussing costs. Ask about prioritizing repairs if you're not able to have everything done at once.

HELPFUL HINTS

Since there is no cut-and-dried way to find a good repair shop, here are some helpful hints and tools you can use to help you make a decision.

ASK QUESTIONS.

Don't ever be afraid to ask questions about your repairs. Being knowledgeable about your car prevents you from becoming a victim.

READ ONLINE REVIEWS.

Local and regional blogs are good sources of information about different repair shops, as are sites that specialize in reviewing local businesses. Like all reviews, take them as a whole rather than on an individual basis.

CONSULT THE BETTER BUSINESS BUREAU (BBB).

Check your local or regional BBB for any issues with the shop you are considering.

CHECK WITH THE MANUFACTURER.

Most car manufacturers have websites and toll-free numbers for registering complaints about service provided to their customers. Contact the manufacturer and ask them about any complaints received about the dealers, or if they have any recommended shops in your area if no dealer is located nearby.

ASK FOR RECOMMENDATIONS.

Ask friends and family members about their mechanic experiences, especially if they drive a car that is similar to yours. Ask them who they recommend, what they liked, and most importantly, what they didn't like about working with a particular shop.

PHONE A FRIEND.

Sometimes talking to a knowledgeable friend or relative can prevent you from making an expensive mistake or help you feel better about the service. Even if they can't fix it, they can be a sounding board for you.

START SMALL.

If you are still not sure about a shop, take them something minor that isn't an emergency to see how they would deal with the repair.

Routine Maintenance Checks

Have you ever seen an old movie where a customer pulls into a gas station and twelve service station employees run out to service the car? Unfortunately, those days are gone—you need to do those basic maintenance tasks yourself to keep your car in top condition.

The following is a quick guide you can use to remember routine maintenance items for your vehicle. Doing these checks may seem like a hassle, but they could save you time and money in the long run. Try to get yourself into some good habits while driving and maintaining your vehicle.

EACH TIME YOU USE YOUR CAR

1. Check the tires for low pressure. If you do see a low tire, look for nails, screws, or other foreign objects that could cause a puncture.
2. Check the outside for new damage.

   This is especially important if you are parked in a public parking area. If someone has damaged your vehicle, take pictures for your insurance company.
3. Look for leaks. After leaving a parking spot, glance back and check for signs of a fresh puddle. When you walk away from your parked car, look for dripping liquid.

EACH TIME YOU GET FUEL

1. Check your fluids. Check the level and condition of your engine oil every time you fill up. The other fluids that have a see-through reservoir like coolant and brake fluid only require a quick glance to see their levels are correct.
2. Inspect belts and hoses.

   Take a look at your belts and hoses to look for abnormal wear. Use caution when touching a hot engine—a visual inspection is sufficient. Also look for leaks around the engine compartment and under the car.
3. Check tire pressure. (Chapter 14)

   This is a good time to check your tires for proper inflation. Frequent checking is the best way to maximize the life of your tires.
4. Clean your windows and wiper blades. (Chapter 15)

   If a squeegee and towels are available, take advantage of them. Keeping your wipers and windshield clean will extend the life of your wipers.

EACH OIL CHANGE

Whether you change your own oil or have it done by someone else, you should perform the fill-up inspections as well as the following:

1. Inspect your tires for abnormal wear and damage. (Chapter 14)
2. Check your fluids. In addition to checking the levels, look at the fluids while the engine is cold, and check their condition.
3. Inspect and lubricate the chassis. (Chapter 13)

   Check for underside problems and issues with the suspension when lubricating the chassis.
4. Inspect the belts and hoses. (Chapter 3)

   With the engine cold, do a thorough inspection of the belts and hoses. Look for signs of leaking around the engine.
5. Check the air filter. (Chapter 5)
6. Check the external lights.

   Turn on your lights and make sure all your lights are working, including your turn signal and brake lights. Also check the small bulb over your license plate.
7. Inspect your battery. (Chapter 11)

   Inspect and clean the battery contacts if needed. Look for damage to the cables.

EVERY OTHER OIL CHANGE

In addition to the previously described checks, add the following to your list:

1. Rotate the tires. (Chapter 14)
2. Check automatic transmission fluid and power steering fluid. (Chapters 4 and 13)

   These are both usually checked while the engine is running.
3. Check the fuel cap. Look for cracking and debris around the seal area.
4. Check the seat belts.

   Check the belts for fraying and damage, and check the retractors to make sure they are working properly.
5. Check weather stripping and seals. (Chapter 15)

   Check the door seals and weather stripping for tears, breaks, or damage.
6. Check computer codes. (Chapter 10)

   The computer should illuminate a dashboard icon if there's a problem, but it doesn't hurt to check for codes periodically.

Ten Tips for Maintaining Your Car

Performing some auto maintenance yourself can save you money. Here are 10 tips to consider if you decide to do your own work. Keep them in mind even if you have someone else work on your car.

1. PUT SAFETY FIRST.

   The most important part of working on a car is your safety. If you plan on working on your own car, you need to invest in a good car jack and safety jack stands. When working on your car, try to find a stable, level, clear area.

   Safety gloves and glasses are also important. Medical exam gloves will keep your hands clean while allowing dexterity. A pair of mechanic's gloves will offer further protection against cuts and heat. Safety glasses are a must. Whenever possible, work on the vehicle when it is cold to prevent burns. Be careful around a running engine; don't wear clothing that can get pulled into the belts.

2. KNOW YOUR LIMITATIONS.

   Be honest with yourself and know if you are capable of performing the job. If you don't have the physical ability to do a job, take it to a pro.

3. KNOW THE JOB AT HAND.

   Research the job at hand before attempting to perform it. Read this book, read your manual, and research it on the internet to find out what types of issues others have run into when trying to do the same job.

4. TAKE ADVANTAGE OF FREE SERVICES.

   Many auto parts stores offer free services that can be a big help to at-home mechanics. Some have a "loan a tool" program that allows you to put down a deposit and borrow the tools for the job. Some will check your codes or change your battery for free. Find out what your local store offers.

5. TAKE PICTURES AND MARK LOCATIONS.

   Take many pictures before and during a project. These photos will help you remember how to put things back together after you take them apart. If you are doing the same thing to both sides of the car, like changing brake pads, do one at a time so you have the other side as a reference. If a piece is aligned in a certain way, mark its location with a dot of paint or correction fluid before you remove the part.

6. **KEEP RECORDS.**
   If you change your own oil, you won't have a sticker in the corner of your windshield telling you when you need another change. Keep a log of your regular maintenance. Many replacement parts come with a warranty; keep a record of the time you installed the part as well as the receipt in case the part fails.

7. **RECYCLE.**
   It is very important to recycle your used car parts and fluids. Find out where you can recycle in your area. Many auto parts stores recycle the basics. For more information on recycling, visit www2.epa.gov/recycle.

8. **HAVE A CONTINGENCY PLAN.**
   Murphy's Law ("anything that can go wrong, will go wrong") is alive and well in auto maintenance. Don't begin a repair without having a contingency plan in place in case something goes wrong. You don't want to find yourself unexpectedly unable to use your vehicle.

9. **ESTABLISH A BUDGET.**
   Establish a budget for both regular maintenance items and for emergency repairs. Having the ability to buy the part at the time the old one fails can save you from more expensive repairs down the road.

10. **KNOW ABOUT YOUR WARRANTIES.**
    Some warranties may require the dealer to perform certain repairs on your car. Items like the catalytic converter carry government-required minimum warranties that may exceed the standard warranties on your car. Some warranties, such as those on a battery, may be prorated because of normal use. Installing aftermarket items on your car may invalidate warranties currently in place. Know your warranties and keep them in your records.

BE PREPARED

IN THIS CHAPTER

What to Carry in Your Car

A roadside emergency can happen at any time, regardless of the age or condition of your car. Being prepared in case of an emergency is a very important part of car ownership. Let's take a look at what you need to carry in your car.

Air Pressure Gauge Checking your air pressure frequently is one of the best ways to avoid an emergency situation.

Flashlight For nighttime emergencies, you need a good flashlight in your car. Be sure to change batteries regularly so the light is ready to go.

Safety Gear Reflective triangles, safety flags, and reflective vests all provide visibility so other drivers know to avoid you. Do not carry safety flares. Flares can be dangerous, and sitting in a hot trunk for a long time can make them unstable.

Tool Kit You can't carry a tool for every situation, but a good, generic tool kit with the correct type of tools (metric or standard) will have you covered for most minor roadside repairs. Look for one that comes in a case for easy storage.

small cables

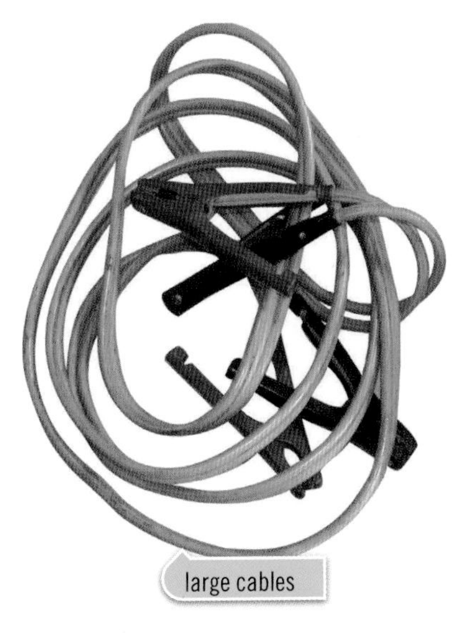

large cables

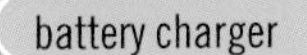

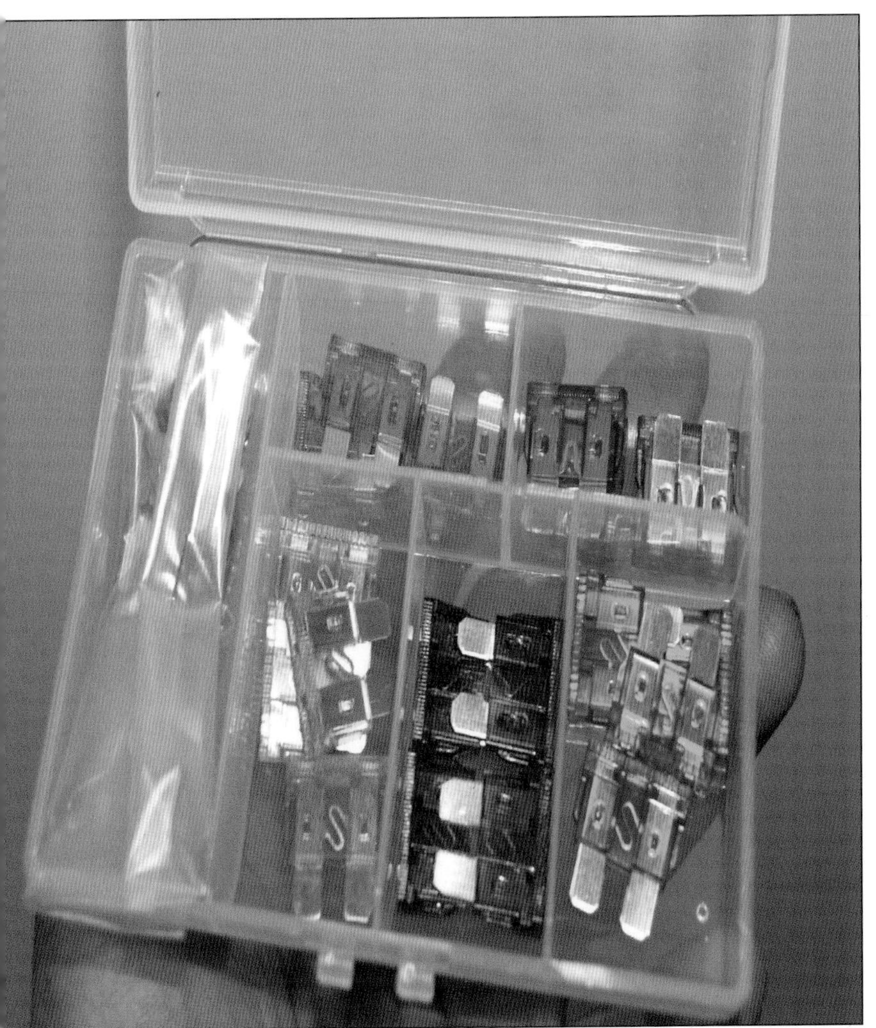

Jumper Cables or Portable Battery Charger Dead batteries are a common problem, even in new cars. Bigger engines require more power to turn them over, and they usually have bigger batteries. The bigger the jumper cable wires, the more power they can transfer. If you have a big engine, get a big set of wires. A small car can get by with a more compact set.

An alternative to jumper cables is a portable battery charger. These devices give you a couple tries at starting your vehicle without having to have another running car. Check the charge regularly so it will be ready to go if you need it.

Spare Fuses One of the more common problems out on the road is a blown fuse. Make sure you match the type of fuses used in your vehicle.

Jack and Wrench New cars come with a jack and tire lug wrench for changing tires. Make sure you know where these are stored in your vehicle.

First Aid Kit Keep some basic first aid supplies on hand to treat the small cuts and burns that can happen when working on a car, as well as other minor injuries.

Glove Box Necessities Your glove box should contain your owner's manual, insurance card and registration, and a hard copy of your emergency contacts and roadside assistance numbers. Consult your owner's manual for proper tire jacking points and the fuse box location, as well as how to use installed vehicle safety features.

Duct Tape Use duct tape for temporary repairs and to keep loose parts from rubbing against each other.

Spare Tire Know where your spare is located, and inspect it regularly. On some vehicles the spare is located under the car, where it is subject to lots of road debris, water, and mud. Make sure the mechanism for removing the spare (bolt or crank system) moves freely. You won't be going very far if your spare is flatter than the one you want to replace.

Fire Extinguisher Not absolutely necessary, but good to have if you need it. If you do carry a fire extinguisher in your vehicle, make sure you get proper training on how to use it.

Emergency Power Cell phone batteries can die, so keep an emergency power source in your car, whether it is an inexpensive phone backup battery or a more sophisticated system for other devices.

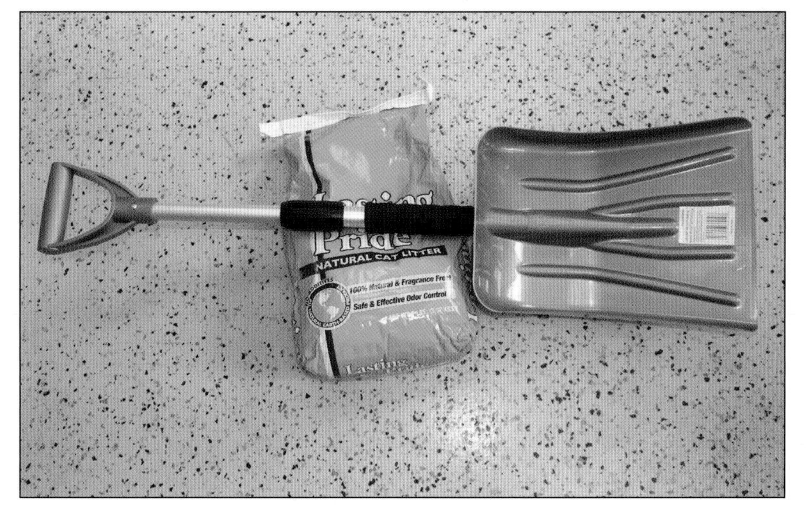

COLD WEATHER EMERGENCY EQUIPMENT

Cold-weather climates require additional equipment to help you deal with ice and snow.

Ice Scraper Don't rely on your car's defroster to melt the ice off your windows.

Cold Weather Kit If you get stuck and help might be awhile, a cold weather emergency kit will keep you protected. This should include a blanket or sleeping bag; drinking water; granola bars or other non-perishable, high-protein food; and hand warmers.

Sand and Shovel Sand or kitty litter can be used as ballast weight in a rear-wheel-drive car, and can be spread in and around tires to provide traction on ice or snow. A compact folding shovel can help you dig out of a snow bank.

Coffee Can, Candles, and Matches A coffee can and tea light candles provide light and enough warmth to melt snow if needed. Be careful burning candles and keep a window open for ventilation.

Snow Tires If you live in a severe climate, consider a set of snow tires. All-season radials are good in mild or moderate cold weather.

Roadside and Emergency Safety

Emergency conditions can occur in a parking lot or while you are driving on the highway. The following are general rules for safely dealing with those conditions and preparing to fix a problem should it arise.

ACCIDENTS

There are many circumstances in which accidents can occur. Here are some general guidelines for minor accidents that bring your vehicle to a halt.

1. REMAIN CALM.
   Even a minor accident can be chaotic, and experienced drivers can make mistakes. Stay calm, don't lose your cool, and keep your conversations with other involved parties to a minimum.

2. SECURE YOURSELF AND YOUR PASSENGERS.
   Check yourself for injuries that may prevent you or your passengers from exiting the vehicle. If you are injured, stay where you are until help arrives.

3. CHECK THE SURROUNDINGS.
   If it was a minor incident and your vehicle will run, move it to a safe location. Before exiting a vehicle after an accident, check for traffic, weather, and accident debris that might not make it safe to exit the car.

4. CALL EMERGENCY PERSONNEL.
   Always call the police, no matter how minor the accident. Some insurance companies insist that you do this for claims.

5. SECURE THE AREA.
   If it is safe to do so, secure the area with your safety triangles or flags. Wear a safety vest so other drivers can see you.

6. CONTACT YOUR INSURANCE AGENT.
   Some companies require you to contact them immediately after an accident. Notify them as soon as possible.

7. EXCHANGE INFORMATION.
   Get the following information from the other driver:

   Full name, address, and phone contacts. If the other driver is not the registered owner of the vehicle, get the contact information for the owner.

   Name of their insurance company and policy number.

   The vehicle's make, model, color, and license plate number.

   Phone numbers and contacts of any witnesses.

8. DOCUMENT THE INCIDENT.
   Note the time and location of the accident and take take pictures of the area for your own records.

ROLLING CONDITION

A rolling condition is a problem that occurs while the car is in motion, causing you to pull over to the side of the road. These could include a flat tire, hitting an animal, severe weather, or any situation that doesn't involve other drivers or people. Once you have assessed the problem, take the following steps.

1. TURN ON YOUR EMERGENCY HAZARD LIGHTS.
   Hazard lights will alert other drivers to your situation and signal them to take precautions.

2. FIND A SAFE SPOT TO PULL OVER.
   Try to get to the safest place your vehicle will take you. The farther you are from traffic, the better.

3. SECURE YOURSELF.
   In most cases, the safest place to be is in your car. If it is safe to get out of your vehicle and you can move to a safer location, do so.

4. SECURE THE AREA.
   If you can do so safely, secure the area by putting safety triangles or flags behind your vehicle to inform other drivers. Raise your hood to show that you need help.

5. ASSESS THE SITUATION.
   Once the area is secured, assess whether you can fix the problem safely on your own. If you can't, contact your roadside assistance or emergency personnel.

6. DOCUMENT THE PROBLEM.
   Note the time and location of the incident and take photos if needed for insurance purposes.

NON-MOVING CONDITION

A non-moving condition occurs when something has happened to your parked car, such as a flat tire, a dead battery, being sideswiped by another vehicle, or locking yourself out of your car. After you have assessed that your car is non-operational, here are some steps you should consider.

1. SECURE YOURSELF.
   Find a safe place to wait for help. In the case of a non-moving condition, waiting in your car is usually safe.

2. SECURE THE AREA IF NEEDED.
   Your car may be causing a danger to you or to others (smoke, broken glass, flammable liquids). If it is safe to approach the car and the hazard, put out safety triangles near the hazard to warn others not to approach.

3. ASSESS THE SURROUNDINGS.
   Check to see if your vehicle is accessible for assistance. Will jumper cables reach from another vehicle? Can you use a jack? If your vehicle can be moved safely, look for a nearby place that would allow for emergency attention and move it.

4. RAISE YOUR HOOD.
   Raising your hood is the universal signal that you have a problem with your car. If it is safe to raise the hood, do so. It notifies others you have a problem and helps emergency response personnel find you. If you have power, turn on your emergency flashers.

Avoiding Lockout

Locking the keys in the car is one of the most common reasons people call for roadside assistance. Although new technology is making it easier to avoid this problem, it's worth taking some preventative measures, particularly if you drive an older vehicle.

TRANSPONDERS VS. FLAT KEYS

Most cars today use a transponder key. The transponder sends a radio signal to the car, which lets the car know it is okay to turn on. A flat, metal key won't start most new vehicles, but they will open the door.

HIDING A KEY

Flat keys are inexpensive and hide well. Having a spare in your wallet or purse is always a good idea, but doesn't work if someone borrows your car and then locks the keys in it, or if you leave your purse in the car, too. Think about a place to hide a key that works best for you.

In a magnetic box. These special boxes have a compartment for a key and are designed to attach to the underside of your vehicle using powerful magnets. Make sure you have a solid, horizontal, metal surface to mount the box. These may not work as well in newer cars that have more plastic on the underside.

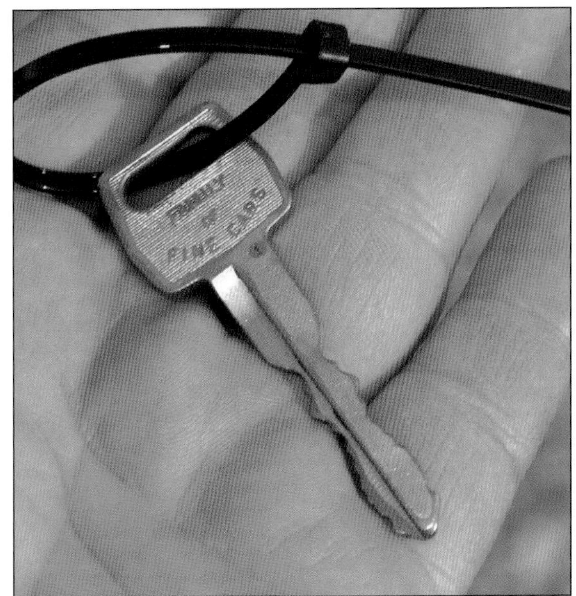

Behind the license plate. Most license plates have a nylon nut behind them for the license plate to screw in to. This keeps the screw from rusting and makes it easy to change plates. It's also a nice place to keep a spare key. You can use a coin to remove the screw and get the stashed key. You may have to drill a small hole in the key to make it fit.

With a small zip tie. You can zip tie a key under the car to a solid metal bolt or hard point. Don't use a fuel or brake line or electrical line.

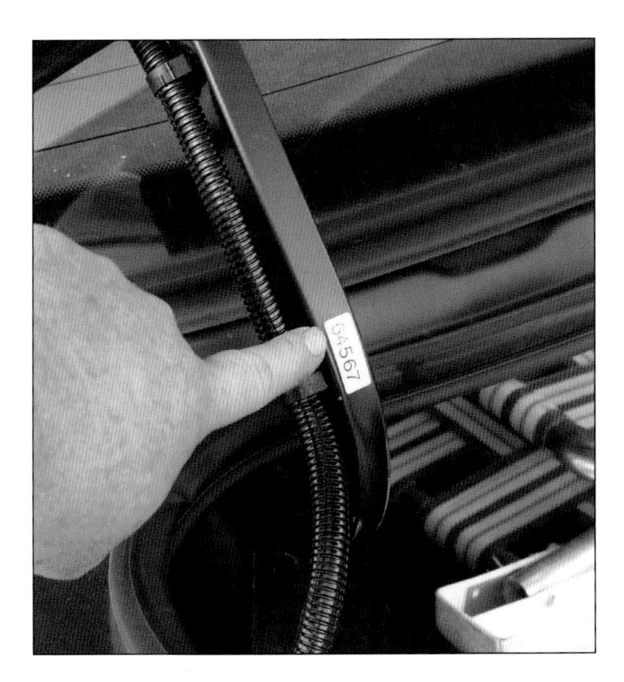

USING KEYPAD CODES

If your car is equipped with an external keypad, the manufacturer usually marks the car with the correct code, in this case on the trunk lid hinge. Write this number down and keep it with you.

EMERGENCY SITUATIONS

Roadside assistance may take an hour or more to get to your location. If you find yourself locked out in an unsafe situation, or if you accidentally lock a pet or child in a car, don't hesitate to call the police. They have tools to get into locked vehicles, and in a worst-case scenario, they can safely break a window to gain access.

How to: Safely Raise a Car

Some maintenance and repair tasks will require you to raise your vehicle in order to access parts under the car. Improperly supporting the car is dangerous, and you need to know how to do it safely. Never use just the floor jack to support the vehicle while working under it—the car can slip off of the jack and crush you.

WHAT YOU NEED

- Jack and jack wrench
- Safety jack stands rated for your car
- Wheel chocks (4)

1. PREPARE THE CAR.
   Park your car on a flat, level surface. Remove any obstacles around the car that may prevent you from freely getting out from under the car. Set the parking brake, even if you are lifting the rear wheels.

2. PLACE WHEEL CHOCKS.
   Place a wheel chock, a block of wood, or other solid object in front of and behind the wheel that is diagonal from your lift point. If you are lifting the front end and starting on the driver's side, chock the passenger side rear wheel (FIGURE A).

3. POSITION THE JACK.
   Consult your owner's manual for the proper place to raise your vehicle. If you are using your car's jack, check for special notches or locators on the jack and car body (FIGURE B).

4. RAISE THE VEHICLE.
   Use the jack wrench to turn the jack and slowly raise the vehicle. The higher you raise a vehicle, the more unstable it can be. Try not to raise the vehicle higher than is needed to safely do the work required, and pay attention to the balance of the car on the jack. It should be stable (FIGURE C).

5. PLACE A SAFETY STAND.
   Once you have the vehicle at a good height, place a safety stand under the car. Be careful not to pinch fuel, exhaust, or electrical lines running under the vehicle. Lower the jack until the safety stand is holding the weight of the vehicle (FIGURE D).

TIPS

- Never use concrete blocks, wood, or anything other than an approved safety jack stand rated for your vehicle's weight to support your car. Blocks can crumble and wood can break.
- Never use a jack on its own to support the vehicle while working under it. Be careful when changing a tire in an emergency if your stands are not available.
- You can use your jack as a backup to the safety stands. If it does not impede you under the car, place the jack in position and raise it until it just touches the vehicle to provide support in case of a slipping safety stand.
- Only raise the car as high as necessary, usually no more than 6 to 8 inches (15 to 20 cm). If you need to raise the vehicle higher, do it in several steps to avoid twisting the car and causing the safety stands to lean as you raise the car.

6. TEST THE SAFETY STAND.
Before getting under the car or raising the other side, give the body of the car a slight push to check the stability of the jack stand. It should not wobble.

7. INSTALL THE SECOND JACK STAND.
Safety stands should be used in pairs, one on each side of either axle. Repeat the process to raise the other side of the vehicle and install a second safety stand. Retest both stands for stability before performing any work underneath.

8. REMOVE THE CAR FROM THE JACK STANDS.
When the work is completed, and with the wheels chocks still in place, raise the car with the jack just enough to remove one safety stand at a time, and lower the vehicle to the ground.

CAR RAMPS

Car ramps are an alternative method for raising your vehicle. You simply drive up on the ramps and you're ready to go. However, car ramps can be hazardous and should be used with caution. The ramps may slip, and driving too far forward or off one side can seriously damage your car. If you have a low-slung car, you can damage the front end of the car if it contacts the ramps.

How to: Change a Tire

Every driver should know how to change a tire. You may find yourself in a situation without coverage or backup, and it's important to get the tire changed as quickly as possible so you can get back on the road safely. Even if you don't change it yourself, you should observe the process to see that it is changed properly.

WHAT YOU NEED

- Spare tire
- Wheel chock (optional)
- Jack and jack wrench
- Lug wrench
- Tools to remove coverings or locking lug nuts

1. SECURE THE AREA.
   If your flat has occurred in an area where other traffic is near, secure the area with your safety warning equipment. Try to have the vehicle parked on a level surface.

2. LOCATE YOUR SPARE TIRE AND TOOLS.
   The spare tire may be stored in the trunk, under the back of the car, or on the outside. Check your owner's manual for the location of the tire, jack, wrench, and jacking instructions (FIGURE A).

3. SET THE PARKING BRAKE.
   Even if one of the back tires is flat, setting the parking brake will hold the other one. Always put the car in "Park" or in gear and set the parking brake before changing the tire (FIGURE B).

4. CHOCK A WHEEL.

If you don't have a wheel chock with you, placing a big rock or other solid object behind and in front of one of the other wheels may help prevent the car from rolling or rocking while you change the tire. If you don't have anything available, you can still work without it (FIGURE C).

5. REMOVE ORNAMENTS.

If your car has hubcaps or other ornaments covering the lug nuts, remove these items. Check your owner's manual to see if your covers require special tools to remove them (FIGURE D).

6. LOOSEN THE LUG NUTS.

The lug nuts should be loosened on the ground, but not removed before you raise the vehicle. If you try to remove them when the car is raised, the tire may spin. Loosen them just enough so they will spin free when you raise the car (FIGURE E).

LOCKING LUG NUTS

Some cars with expensive wheels may have lug nuts that require a special tool to remove them. If your car has locking lug nuts, insert the tool into the lug wrench and loosen the lug nut.

7. RAISE THE VEHICLE.
   Your owner's manual or sticker will show the proper location to raise your car safely. Put the jack in place and raise the vehicle until the tire is just off of the road surface. Make sure you raise the car high enough to install the spare, but be careful: the higher you raise the car, the less stable it becomes (FIGURE F).

8. REMOVE THE FLAT TIRE.
   Once the vehicle is raised, remove the lug nuts and remove the flat tire by pulling it away from the vehicle.

9. INSTALL THE SPARE TIRE.
   Install the spare tire, paying attention to how it mounts to the car. Temporary spares may have a different look than the original wheel, so make sure you mount the spare with the correct side pointing outward. Install the lug nuts and snug them to the wheel (do not tighten fully yet).

USING A SPARE TIRE

There's a big difference between using a full-size spare and using a temporary spare. A temporary spare is intended only to get you to a place where you can get your tire repaired or replaced. Never drive on a temporary spare longer than necessary.

If your spare does not match your other tires (it's older, or a different brand), have your flat tire replaced or repaired soon. Mismatched tires can cause performance issues.

10. **LOWER THE VEHICLE.**
 Make sure there are no tools or other items under the vehicle and slowly lower the car down to the ground

11. **TIGHTEN THE LUG NUTS.**
 When you tighten the lug nuts, use an alternating pattern to tighten them evenly. With four lug wheels, tighten the lug nuts in a crossing pattern, and with a five-lug wheel, use a star pattern (FIGURE G). After driving on the spare, check the lug nuts to make sure they have not loosened.

12. **SECURE LOOSE PARTS.**
 Stow the flat tire, ornaments, and hubcaps until the tire can be repaired or replaced.

How to: Jump-Start a Battery

At some point, most drivers will encounter a situation where they need to jump-start a dead battery. This can be done with a booster battery from another vehicle or with a portable power pack. Whichever method you use, it's important to do it correctly.

WHAT YOU NEED

- Jumper cables or a power pack
- Spare battery or running vehicle
- Battery terminal brush (optional)

JUMP-STARTING WITH A BOOSTER BATTERY

The most common way of jump-starting a battery is by connecting it to the battery of another vehicle using jumper cables. The charging power of the running vehicle gives the dead battery enough of a boost to turn over the engine.

1. LOCATE THE BATTERIES.
   Locate the batteries on both vehicles. They are usually mounted under the hood. If it isn't under the hood, check the owner's manual for the location. The location of the batteries will determine how to orient the vehicles (FIGURE A).

2. PARK THE BOOSTER VEHICLE.
   Park the booster vehicle next to the disabled vehicle so that the batteries are as close as possible. Often this means parking the cars nose-to-nose. Make sure the jumper cables are long enough to reach from one battery to the other.

REMOTE BATTERY LOCATION

Some batteries are located in a spot that is difficult to access, such as the spare tire area of the trunk. Most manufacturers provide a safe point to jump-start these vehicles. Check your owner's manual for the exact location on your car.

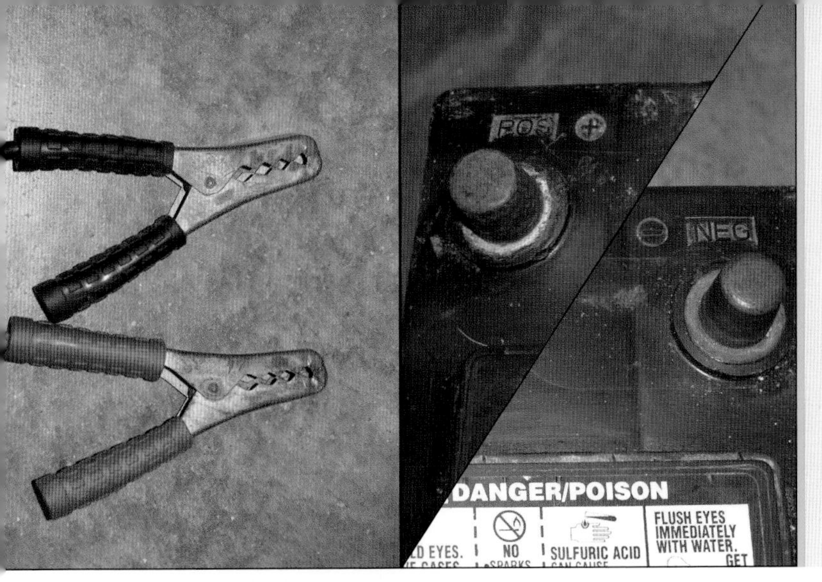

IDENTIFYING CABLE ENDS AND TERMINAL POSTS

Most jumper cables are color-coded. Red is used for positive voltage and black is used for the grounding cable.

The positive terminal post on a battery often has a red cap and is labeled with a plus sign (+). The negative terminal post often has a black cap and is labeled with a minus sign (-).

3. TURN OFF ALL ELECTRONIC DEVICES IN BOTH CARS.
   Power surges can occur during jump-starting, ruining electronic components. Unplug all personal devices and turn off the radios and other devices in both vehicles.

4. TURN ON THE HEATER FANS.
   The heater fan uses a lot of power, and if you do have a surge, it will be able to absorb the power spike. Turn the fan on in both cars.

5. CLEAN THE TERMINALS ON BOTH BATTERIES.
   Excess corrosion will prevent the jumper cables from making good contact with the terminals. If you have a battery brush or stiff wire brush, clean the battery terminals to allow for a good connection (FIGURE B).

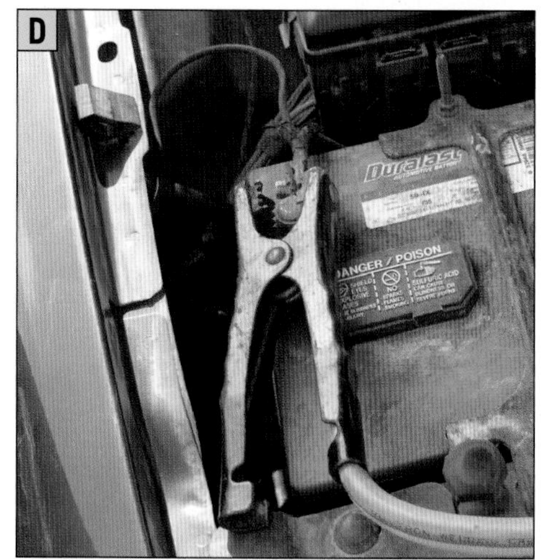

6. **HOOK UP THE POSITIVE (RED) CABLE.**
   Pull the terminal covers out of the way and hook one red clamp to the positive post of the disabled battery, then hook the other red clamp to the positive post on the booster battery. Make sure the clamps are secure and do not touch the metal chassis of the car (FIGURE C).

7. **HOOK UP THE GROUND (BLACK) CABLE TO THE BOOSTER BATTERY.**
   Pull any terminal caps back and place a black clamp on the negative post of the booster battery. Be sure the cables and clamps are not near moving parts (FIGURE D).

8. **ATTACH THE GROUND (BLACK) CLAMP TO THE DISABLED VEHICLE.**
   The opposite end of the black cable needs to be grounded. Do not connect it to the battery negative terminal. Instead, find a good, solid spot on the chassis to make the clamp connection. Here we used the bolt from the front suspension for our ground (FIGURE E).

9. **START THE BOOSTER VEHICLE.**
   Start the booster vehicle and run it at a moderate idle speed. Allow the booster car to charge the disabled battery for a couple of minutes before trying to start the disabled vehicle.

CAUTION

Car batteries contain acids and gasses that can be released or explode when the battery is shorted or overcharged. Stand back if you are outside a vehicle while it is being jump-started. If you hear hissing, or the battery is getting very hot, or you see bubbling around the top covers, stop the process immediately.

10. START THE DISABLED VEHICLE.
With the booster vehicle still running, try to start the disabled vehicle. You may have to try several times before it starts. Once it starts, keep both vehicles running for a few minutes.

11. REMOVE THE BOOSTER CABLES.
Take the booster cables off in the reverse order of putting them on. Start with the disabled vehicle's negative clamp. Once you disconnect the cable, wait a couple of seconds to see if the car stops running. If it does, you may have a charging problem. Remove the negative clamp from the booster battery and then remove the positive clamps, starting with the disabled battery.

SPECIALTY BATTERIES

Some specialty batteries may require different recharging methods. Absorbent Glass Mat (AGM) batteries are great, but if the charge falls under 10.5 volts, they can't be recharged by basic battery chargers.

New technologies like gel batteries and deep cycle batteries also may require special charging. Check with the manufacturers for more information.

CHARGING WITH A PORTABLE CHARGER

Charging with a portable charging device is very similar to using a booster battery. Connect the positive clamp to the battery post and the negative clamp to a good metallic ground point on the car. Follow the instructions on your portable charger and remove after the vehicle is running. Make sure you recharge your portable unit so it is ready for use the next time.

How to: Check and Change a Fuse

A blown fuse is one of the more common roadside problems. The result of a blown fuse can be major (the car stops running) or minor (the radio stops working). Here's how to check for a bad fuse and how to replace it if necessary.

WHAT YOU NEED

- Circuit tester or multimeter
- Spare fuses (the type used in your car)
- Fuse puller or long-nose pliers
- Owner's manual

A

B

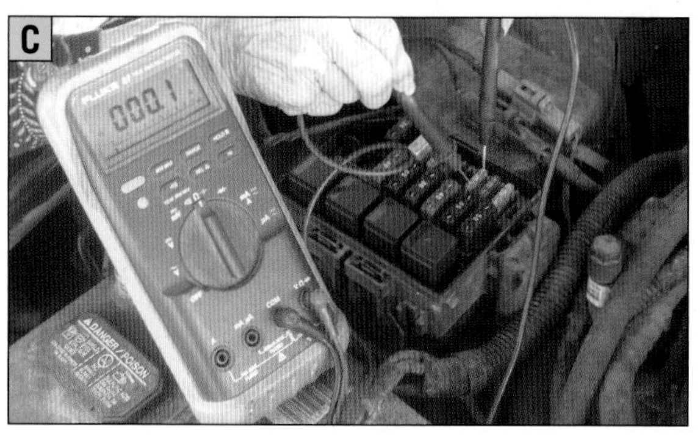
C

CHECKING FOR A BLOWN FUSE

1. LOCATE THE FUSE BOX AND REMOVE COVER.
   The location of the fuse box varies greatly, but they can usually be found in the engine compartment or in the dashboard. Depending on your vehicle, you may have more than one fuse box. Check your owner's manual for the location of all fuse boxes and remove any protective covers (FIGURE A).

2. TEST THE FUSES.
   Place the metal contact or tips of your tester to the exposed metal parts of the fuse. If you make good contact, your tester will light up. If it doesn't, check it a couple of times and check another fuse to make sure you are making good contact. If your tester doesn't light up, you have a bad fuse (FIGURE B).

   If you are using a meter, turn the meter setting to "continuity" or "resistance," usually marked with an omega symbol (Ω). If the fuse is good, the meter will show almost no resistance or zero resistance, and if the fuse is bad, it will not make a connection in your meter (FIGURE C).

   Check all of the fuses for continuity, and note any that are bad.

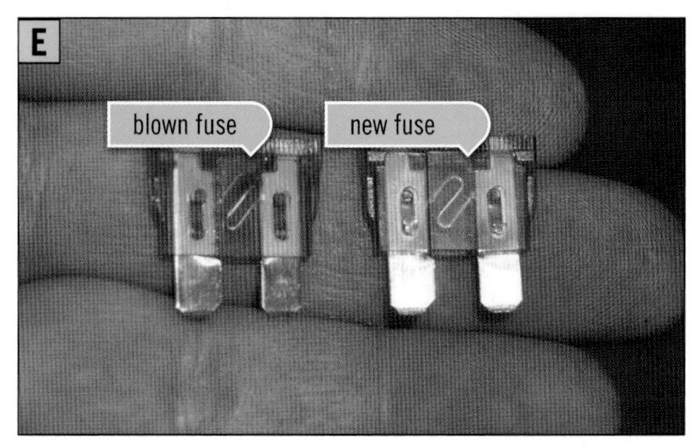

CHANGING A FUSE

1. **PULL THE FUSE.**
   Using either your fuse puller (ours was built into the tester) or a pair of pliers, gently pull the fuse from its pocket (FIGURE D).

2. **INSPECT THE FUSE.**
   If the fuse is blown, the metal in between the two power posts will be melted and will not be connected. You may be able to see a broken or burned spot in the metal (FIGURE E).

3. **INSTALL THE NEW FUSE.**
   Using a fuse of the same size and color (the color is specific to how much power the fuse can protect), press the new fuse into the socket. If the fuse is in a difficult place, you can use your fuse tool or pliers to gently install it. Push it all the way in to make good contact (FIGURE F).

4. **CHECK THE NEW FUSE.**
   Re-check the new fuse to make sure it is good, and then try the electronics that were failing to see if that solved the problem. If you continue to blow circuits, you need to have a professional look at your car.

FUSE TESTERS

You need a continuity tester that will test a circuit with the power off. You can use an expensive multimeter or a simple circuit tester that may be included in your box of fuses. Either will do the job just fine. Your checker does need to be self-powered.

THE ENGINE

IN THIS CHAPTER

How the Engine Works

The engine in your car is doing much more than just making your car go forward. It generates electricity to run the lights and computer, it creates a vacuum to run systems like power brakes, and it produces power to run pumps that power systems such as the air conditioning and power steering.

Most engines—internal combustion, rotary, or otherwise—work off of the same principle. Take fuel, mix it with air, compress it, and then ignite it to create an explosion that releases energy. The engine turns this energy into rotational movement, which runs all of your car's devices.

The primary engine consists of a series of combustion chambers called cylinders, and a system of parts that convert the explosions into motion called the *rotating assembly*. Because most of the parts in an engine are metal, the engine uses an oiling system that prevents friction damage and enables the parts to move freely. Here are some of the major components that make up your engine.

Cylinder or Engine Block The cylinder block can have between 1 and 16 cylinders. Each cylinder is used to create the combustion. The cylinder block is usually made of aluminum or iron, and is designed to hold up to the pressures created when the fuel is detonated. Depending on your engine, it can compress the air and fuel from 8 to 20 times its normal pressure. The engine block holds all the rotating components and has passages for oil and coolant to flow through it. The cylinders may be in a line, in a "V" shape, or opposing each other.

Cylinder Head Your engine may have one or more cylinder heads. They are used to cap off the cylinder so the fuel and air can be compressed. The cylinder head has ports that allow air in and out of the cylinder.

Piston and Connecting Rod The piston makes up the bottom of the cylinder. It is pushed up to compress the fuel to burn it, and when the explosion occurs, it is pushed down to turn the crankshaft. The connecting rod connects the piston to the crankshaft.

Crankshaft The crankshaft is turned by the motion of the pistons. The crankshaft is connected to the drivetrain, which moves the wheels, and the external accessories like the alternator, coolant pump, and air conditioner.

Valves The valves regulate the flow of air and fuel into the engine. When the valves are open, they let the fresh air and fuel into the cylinder and allow the burnt gases to flow out of the exhaust system. When the valves are closed, they allow compression from the piston to occur.

Camshaft and Solenoids The engine needs to time when the valves are open and closed. A camshaft uses egg-shaped metal lobes to move the valves open and closed in time with the crankshaft. Some engines use electric solenoids timed by the computer to open and close the valves.

Oil Pump Oil is pumped through the engine to prevent the metal parts from seizing. Because the oil picks up contaminants from the combustion process, as well as metal particles from the motor, it needs to be changed on a regular basis.

Timing and Accessory Belts The engine needs to keep the camshaft and crankshaft working together, so it uses a timing belt, chain, or gears to keep them in sync. The accessories, such as the alternator and A/C compressor, are also run off of flexible belts. Belts and chains will wear out over time, so they need to be replaced at regular intervals.

Horsepower and torque can be measured by using a machine called a ***dynamometer***. The car runs against rollers that apply resistance to the drive wheels (load), and the machine measures how much the car can spin the rollers with a known resistance. Performance cars can be tuned to optimize horsepower and torque based on the readings from the dynamometer and information from the car's computer.

HORSEPOWER, TORQUE, AND LOAD

The average car only needs about 15 to 20 horsepower to maintain 60 miles per hour, so why do we have engines with horsepower measured in the hundreds? Let's talk about horsepower, torque, and load.

Horsepower Horsepower measures how much work your engine can do. The more horsepower, the more work your engine is capable of. While it only takes 15 to 20 horsepower for an average car to maintain a speed of 60 miles per hour, it takes more work to accelerate, go up hills, and run items like the alternator and air conditioner.

Torque Torque refers to the ability of the car to turn something, like the wheels. When your car is standing still, it takes more torque to overcome anything that's keeping your car from moving. The more torque your car can generate, the easier it is to make the moving parts rotate.

Load Load is the amount of work the engine is being asked to produce. Energy use, such as turning the air compressor, and driving conditions, such as going up hills, put demands on the engine that it must overcome with horsepower and torque.

While regular engine-powered cars need additional horsepower and torque as they drive to overcome load, hybrids can run the engine at a constant rate. This is because the electric motors are creating the primary horsepower and torque, which saves fuel.

Common Engine Problems

The engine makes power by compressing and igniting fuel and air, creates vacuum by pulling the air and fuel into the engine, and runs the accessories by connecting them to the crankshaft. Problems with any of these functions could be due to an issue with the engine itself or with one of the other systems in the car.

"CHECK ENGINE" LIGHT COMES ON

The "check engine" light comes on when the car's computer finds something off with the efficiency of the motor. How the fuel gets in, how well the engine is burning it, how good the spark is, and what the exhaust gases are like can all contribute to the way the engine performs. When the light comes on due to something in the motor, it usually is related to the engine's ability to close off the cylinder and make compression, or it may be due to the engine overheating.

If your "check engine" light comes on, follow the instructions in Chapter 10 and determine the cause. If it is an internal engine problem, you'll need to take it to a professional and have it checked.

LOSS OF COMPRESSION

The engine relies on three things to keep pressure inside the cylinder: valves, which sit hard against the cylinder head; gaskets, which are located between the head and block; and piston rings, which expand between the cylinder wall and the piston. If one of these fails, your engine will have a loss of power or stop running.

If a valve bends, it won't be able to seat against the head; if a gasket fails, it will release pressure; and if the rings fail, it will force pressure into the bottom of the engine and cause the engine to falter. When one or more cylinders start to lose compression, the effect is loss of power, vibrations, and stalling of the motor.

A failure in any of these parts can also cause oil or coolant to enter the cylinder, and that will cause engine problems, too. A professional can run diagnostics on your engine, and will have a tool to check engine compression. Checking the compression in the cylinders tells your mechanic a lot about what is happening inside the engine without taking it apart.

ENGINE TIMING PROBLEMS

The engine uses a belt, chain, or gears to keep the crankshaft and camshaft working together. Belts and timing chains can stretch or slip over time, causing the valves to stop opening and closing when they should.

You may hear your mechanic talk about "interference" motors. Interference motors run much tighter cycles than non-interference motors, and if the rotating parts get out of sync, they can touch each other and break things. A valve touching the pistons is the biggest problem, and it is expensive to fix. Non-interference motors really have to slip a lot before damage occurs.

Changing a timing belt or chain is best handled by an experienced mechanic, because accessing the belt requires disassembling the engine.

OVERHEATING

When the motor overheats, metal expands more than it should. This means the computer has to adjust the air intake to compensate, which can cause premature wear and damage to the engine. The primary culprit is the cooling system. Engine oil also helps cool the engine and its parts, and if there is a problem with the oil, it can cause an overheating issue. If the engine starts to get hot, the first sign is the temperature gauge on the dash moving out of the normal range. You may also see a release of coolant (steam under the hood) and seizing of the motor can occur.

FLUID LEAKS

Fluid leaks can occur when a gasket, which is designed to keep the oil, coolant, and fuel contained, fails. The first indication of a leak is usually a puddle in your driveway or your fluid levels being lower than they should be.

Internal leaks, like compression problems, can cause the engine to start running poorly. Oil or water in the cylinder will cause poor combustion, which you will see in the form of smoke coming out of the exhaust pipe.

Your ability to fix a gasket really depends on where it is and how accessible it is. Just remember—that small drip can turn into a big leak very quickly.

WORN OR BROKEN BELTS

Like the timing belt, the accessory belts will stretch and wear out over time. On some cars it is easy to change the belt or belts. Your ability to do so depends on how accessible the belts are. Covers, fans, and other accessories may prevent you from changing it yourself without special tools.

More and more manufacturers are putting covers and protective shields on cars, making it difficult to inspect the belts. If it is difficult to properly inspect the belts on your car, have a professional check them at the manufacturer's recommended intervals.

VACUUM LEAKS

The engine creates a vacuum by drawing in air when the piston moves down the cylinder. When a vacuum leak occurs, it can change the amount of air that is mixed with the fuel, and this will cause the engine to run poorly. It may idle too high or run rough.

There are lots of vacuum tubes on a modern motor, and when one breaks, you may hear a hissing or whistling sound coming from the break. Fixing a vacuum leak may be as simple as reconnecting a line, but if the line is broken, it will need to be replaced.

A repair shop can use a computer to help find a leak. Sometimes they are located in very hard-to-reach places, and you need to remove a lot of equipment to get to the leak. It's best to leave this repair to a professional.

How to: Check and Add Engine Oil

Checking your engine oil should be a regular routine. Most manufacturers recommend checking it every couple of hundred miles of driving, or each time you get fuel. For most vehicles, you check your engine oil with a dipstick that reaches down into the oil pan and dips into the oil reservoir.

WHAT YOU NEED

- Clean rag or paper towel
- Engine oil (if needed)
- Funnel (optional)

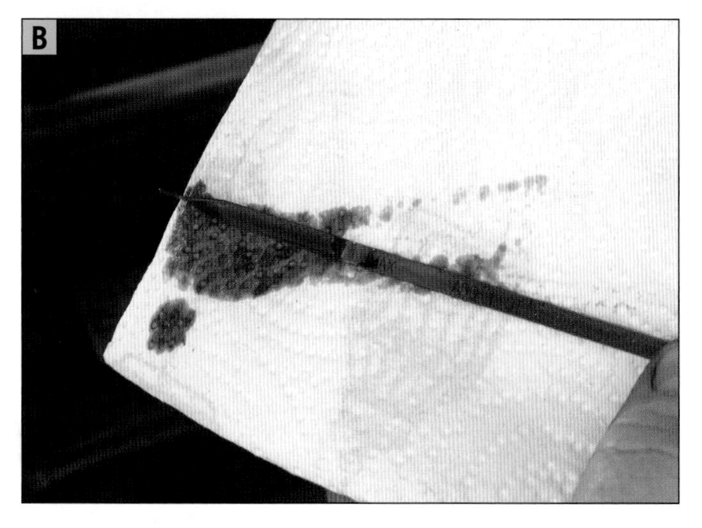

CHECKING ENGINE OIL

1. TURN OFF THE ENGINE.
   You can check the engine oil at any time, but the best is at least five minutes after it has been running and is warm but not hot. Be careful not to burn yourself while checking a hot engine.

2. LOCATE THE OIL DIPSTICK.
   Most manufacturers are very good at marking the engine oil dipstick for identification. On this car it is bright yellow and right up front, but it may look different on your car (FIGURE A).

3. PULL AND INSPECT THE DIPSTICK.
   Remove the dipstick from the engine and inspect it for anything that doesn't look like oil (FIGURE B).

WHAT TO LOOK FOR

Heavily Contaminated Oil It's normal for oil to turn black from the soot of the engine. What you need to look for is grit and sludge in the oil. This will make the oil look thicker and heavier than when you put it in. Clean oil has a brown or dark brown color and has the consistency of syrup.

White, Milky Foam A whitish foam just above the oil level may indicate that antifreeze or condensation has been mixing into the oil. If the dipstick hasn't been fully inserted in the tube, moisture can run down the dipstick and cause this. If it is heavy or persistent, you need to have the engine inspected.

Metal Particles If you see shiny metal flakes in the oil, you may have an engine problem.

If you find one of these problems, change your oil and recheck the problem. If the problem doesn't go away, take your car to a pro and have it checked.

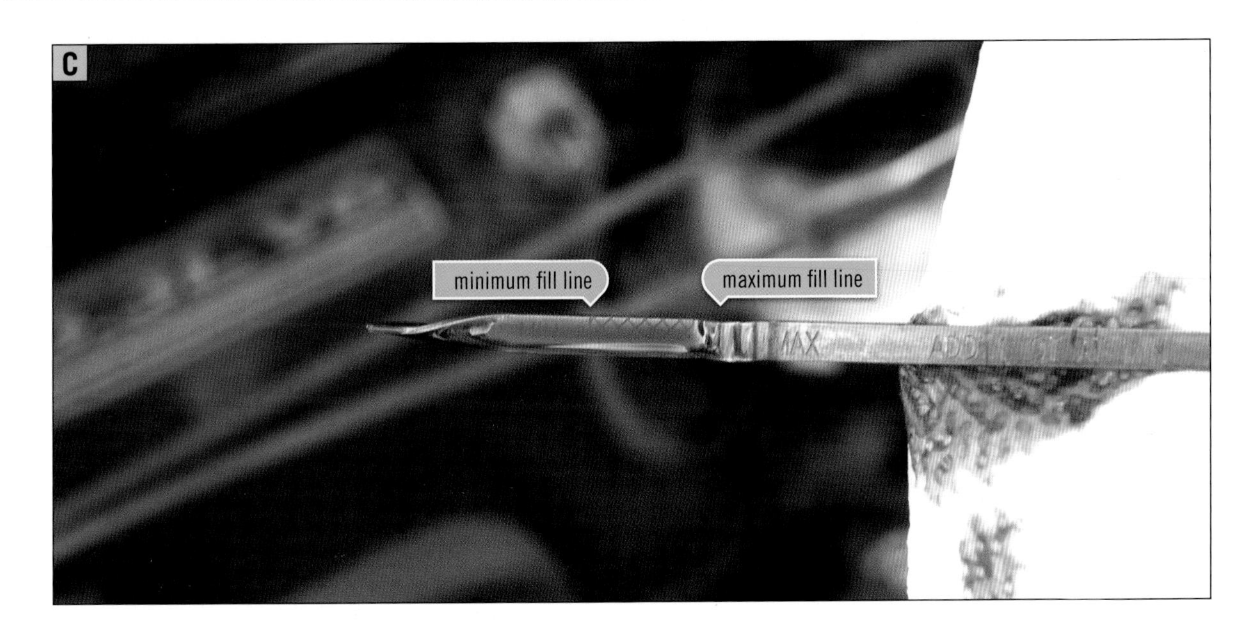

4. WIPE AND RE-INSERT THE DIPSTICK.
   Since the engine slings oil onto the dipstick when running, use a clean rag or paper towel to wipe the dipstick and re-insert it all the way back into the engine.

5. INSPECT THE ENGINE OIL LEVEL.
   Pull the dipstick back out to get a fresh reading. Check that the level falls within the minimum and maximum levels marked on the dipstick. Ideally, the level should be at but not over the maximum level (FIGURE C).

ADDING ENGINE OIL

If your engine is healthy and you get regular oil changes, your oil level shouldn't get low enough between changes that you need to add oil. If it is consistently low, it usually means it is leaking or burning oil. A very small leak is nothing to be concerned with, but all the more reason to keep an eye on the oil level. If you need to add oil yourself, here's how to do it. Be sure to use the type of oil recommended by your manufacturer.

1. LOCATE THE OIL FILL LOCATION.
   The oil fill is usually on top the engine in one of the valve covers. If you can't find it, check the owner's manual for its location (FIGURE D).

2. POUR IN THE OIL.
   Use a funnel to prevent drips and splatters that could be mistaken for a leak (FIGURE E).

3. CHECK THE OIL LEVEL.

Your owner's manual or the dipstick may tell you how much oil is between the minimum and maximum levels. Wait a couple of minutes for the new oil to run down into the oil pan and check the engine dipstick again. Repeat the fill process until the level has reached the maximum line (FIGURE F).

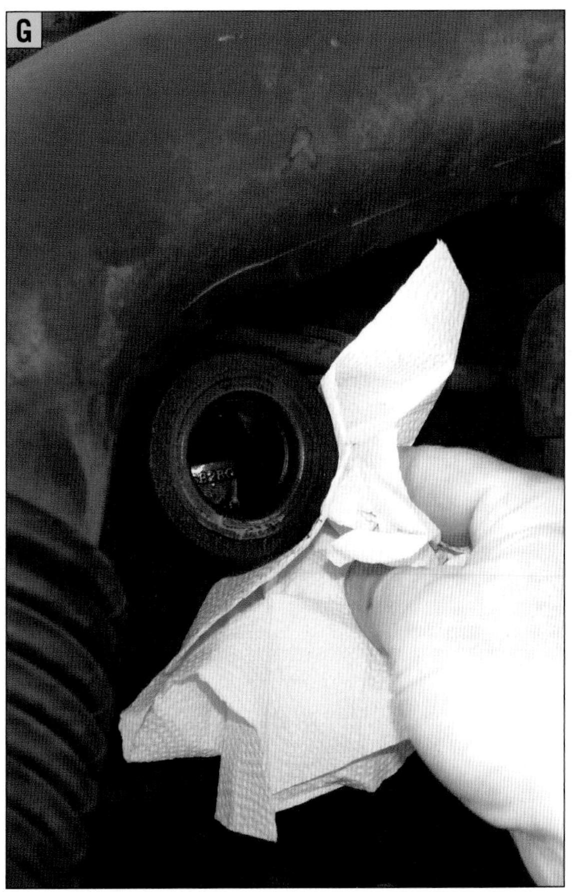

4. CLEAN AND REPLACE CAP.

Make sure you clean any spilled oil from your motor (be careful on a hot engine) and replace the cap (FIGURE G).

OIL TYPES

So, your mechanic has recommended that you use a synthetic oil. They say it's better for your car, but it costs a lot more. What is the difference between conventional and synthetic oils?

Conventional oil is a petroleum product, which has to be refined. Aftermarket engine parts manufacturers may recommend conventional oil when installing their parts. Conventional oil is less expensive than synthetic oil, but must be changed more frequently, usually every 3,000 to 6,000 miles.

Synthetic oil is made in a lab, which results in a more stable and, at times, better-performing product. Because synthetics can stand higher heat, some high-performance manufacturers recommend them. Synthetics can go 10,000 to 12,000 miles before they need to be changed, but since they cost three to four times more than conventional oils, the cost savings may be negligible.

Oil makers may also make "blended oil" which use both conventional and synthetic types to get the advantages of both, but fall between each in performance and ability.

Both conventional and synthetic oils have to meet a government standard, even the inexpensive generic brands. Car manufacturers also set standards for engine oil based on how the engine is made. Engine oil is rated by its weight (a "W" number, like 30W). Thinner oils have lower numbers than thicker oils. Most new cars are designed to use thinner oils, which take less energy to pump, saving gas.

If you're using the manufacturer's recommended grade of oil, your car will operate just fine. A mechanic may recommend synthetic if you aren't getting your oil changed frequently enough or if you do some high-performance driving. If you change your oil on a regular basis, conventional oil is sufficient. Consider a synthetic if you're forgetful, have a lead foot, or like the added protection.

CARS WITHOUT DIPSTICKS

Some new cars do not come with an engine oil dipstick. Instead, they have a sensor that measures the oil level and lets you know if the car needs oil. If your car has this system, it is very important that you change your oil at the recommended intervals and check regularly for leaks.

READING YOUR OIL GAUGE

The oil light or gauge measures the oil pressure as oil is pumped through the engine. Too little pressure means the oil may not be reaching the parts to lubricate them, and too much pressure may indicate a blockage. Both are very bad for your engine.

Some cars use a light, and some cars use a gauge to tell you about the oil pressure. If you see the light come on, or if you see the gauge go up too high or too low, bring your car to a stop as soon as safely possible and check the level of your engine oil.

Do not drive a car with low oil pressure; you will damage the engine. If the engine oil is very low, refill the engine oil and re-check the gauge while the engine is running. If the problem goes away, check for leaks or oil-burning smoke from the exhaust. Have the car checked by a professional as soon as possible. If you find any other problems, do not drive the car, have it towed to a repair garage immediately.

How to: Change the Engine Oil

Changing your own engine oil is a good way to save money, but before you begin, check local regulations to make sure it's legal for you to do it yourself. Recycling oil is a must, and items like the used filter may require proper disposal. Your local auto parts dealer will have information about regulations in your area and may be able to recycle used oil, too. Finally, keep in mind that changing oil is a messy job and your filter may be difficult to access. Be prepared to get dirty.

WHAT YOU NEED

- Oil recommended by your manufacturer
- Oil filter appropriate for your car
- Paper towels
- Oil pan
- Tool to remove the oil drain plug (usually a wrench)
- Jack and jack stands (if needed)
- Funnel (optional)
- Oil filter wrench (if needed)

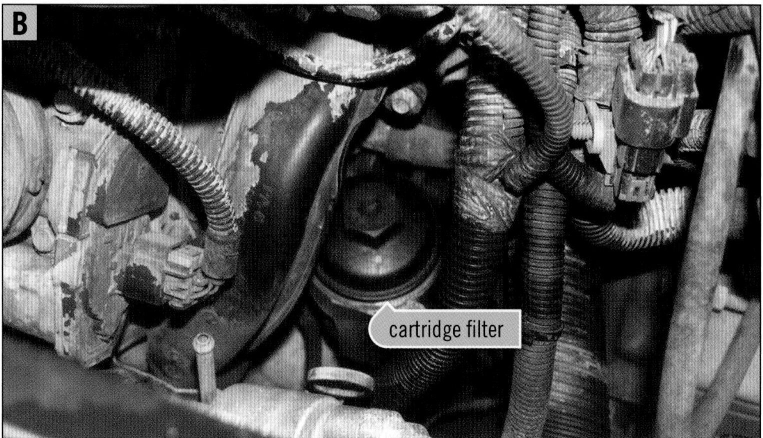

1. LOCATE THE OIL FILTER.
   With the engine cold, locate the oil filter. Make sure you can access it easily and take note of the style of filter your car uses. There are several filter types; the most common are cartridge and spin-on filters.

   This spin-on filter is located on the driver's side of the motor, down by the front steering. Removing it requires adjusting the position of the steering linkage and crawling under the car (FIGURE A).

   This cartridge filter is located up high and tucked under the engine's exhaust manifold (FIGURE B).

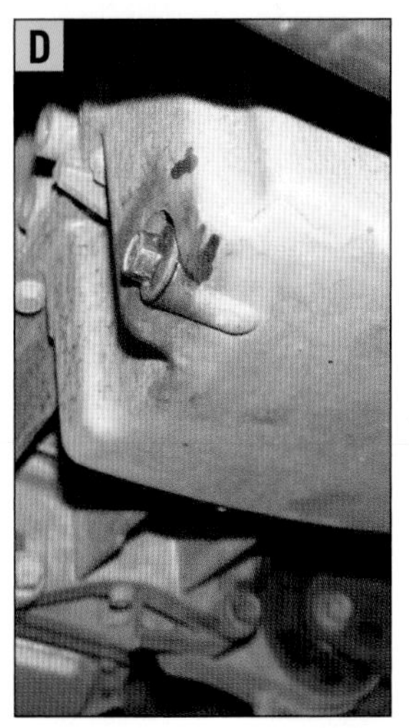

2. PREP THE CAR.
   Find a safe, level place to change your oil. Make sure your car is in "park" or in gear and set the emergency parking brake to prevent the car from moving. If your car is low to the ground, you may need to raise the car up to gain access to the drain plug and filter (FIGURE C).

3. LOCATE THE DRAIN PLUG.
   The drain plug is located at the lowest point on the engine. It usually looks like a bolt (FIGURE D).

4. REMOVE THE DRAIN PLUG AND DRAIN THE OLD OIL.
   Place your oil drain pan under the plug and remove the plug with your wrench. Be careful: the oil will drain fast and can splatter easily. Adjust the position of the pan if needed as the flow of oil slows. Allow the oil to drain completely (FIGURE E).

5. INSPECT THE DRAIN PLUG.
   Before you reinstall the plug, inspect it for damage or contaminants. Clean the plug and threads with a paper towel before re-installing.

6. REINSTALL THE DRAIN PLUG.
   Reinstall the drain plug and tighten. Be careful not to overtighten the plug and strip the threads (FIGURE F).

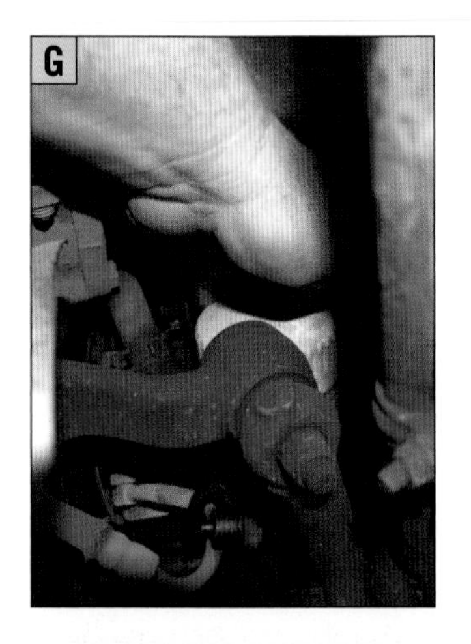

7. REMOVE THE OIL FILTER.
   Removing the oil filter can be tricky. You may be able to do it by hand, or you may need to use a special oil filter wrench. Place your oil drain pan under the filter in case of leaks.

   This spin-on filter is awkwardly positioned, but can be removed by hand (FIGURE G).

   Taking out this cartridge filter requires maneuvering a wrench under some wiring and removing the top cap of the filter canister (FIGURE H).

8. INSPECT THE FILTER.
   Once you have the filter off and away from the car, look it over. On a spin-on filter, you can't see the inside, but you can check the opening for metal shavings or debris from the engine. Check that the gasket on the filter is intact and hasn't stuck to the engine (FIGURE I).

   On this cartridge filter, you can see the filter element and check it for metal shavings or large contaminants. This cartridge has an O-ring gasket on the bottom and on the threads. Make sure the old gaskets come off with the filter and are not stuck to the engine (FIGURE J).

USING A FILTER WRENCH

If your filter will not come off by hand, you may need to use a special oil filter wrench. These are available in several different styles, including the band style (shown). If you try a wrench and you still can't get the filter off, you can put the old engine oil back in the motor, seal it up, and take it to a professional.

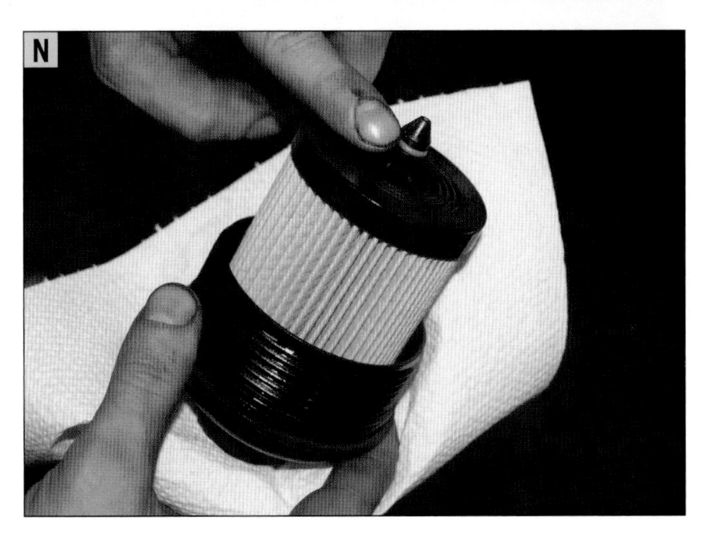

9. CLEAN THE GASKET SURFACES.

The gasket surfaces need to be cleaned before installing the new filter.

On a spin-on filter, use paper towels to wipe off the flat surface on the engine where the filter gasket touches the engine (FIGURE K).

On a cartridge filter, check the bottom of the canister for contaminants and use a paper towel to wipe them out. Be careful not to leave any paper towel or any other foreign object in the canister. Use a paper towel to clean the threads on the top of the canister (FIGURE L).

10. LUBRICATE THE NEW GASKET.

For a spin-on filter, smear a small amount of new engine oil on the entire mating surface of the new gasket before re-installing it on the engine. This will help seal the filter and make it easier to remove on the next oil change (FIGURE M).

For a cartridge filter, snap the new filter into the top cap and use a small amount of oil to coat all the gaskets. Most cartridge-type filters will come with a new O-ring gasket for the cap. Install the new O-ring and coat it with some clean oil to help seal the gasket and make removal easier next time (FIGURE N).

CHOOSING AN OIL FILTER

You're at the auto parts store—which oil filter should you buy? There are many choices. Here are some guidelines to keep in mind.

1. Changing your oil on a regular basis is more important than what type of filter you use.
2. Premium price does not always mean a premium product. However, if you change your filter regularly, any well-known brand of filter will work just fine.
3. Avoid very low cost and no-name filters.

11. INSTALL THE NEW FILTER.

A spin-on filter is designed to be tightened by hand. To install a spin-on filter properly, screw the filter on the mount until you feel the gasket touch the mounting surface. At this point, turn the filter approximately three-quarters of a turn. Check the packaging for specific instructions.

For a cartridge filter, screw the filter cap on by hand until the gasket fully seats and then snug the cap. Do not overtighten; you can tear or ruin the gasket.

12. POUR IN THE NEW OIL.

Once the new filter and plug are back in the engine, remove the oil cap and pour in the recommended amount of motor oil. Using a funnel helps to prevent spills (FIGURE O). Use paper towels to wipe up any spilled oil and replace the oil cap.

13. CHECK FOR LEAKS.

Before starting the engine, check under the car for any drips from the drain plug or from the oil filter. Turn on the car and let it run for a minute or two and check under the car for leaks.

14. REMOVE THE CAR FROM JACKS, IF NEEDED.

Once you are sure you don't have to get back under the car, you can lower the car off the safety jacks.

15. RESET YOUR COMPUTER IF NECESSARY.

Some vehicles come with a vehicle data center that includes a feature that tells you when to change your oil. The step for resetting this feature should be in your owner's manual.

If you don't have a reminder, note the mileage and date when you performed the oil change and keep that information in your car.

TIP

Some people recommend that you put oil in the new filter before installing it so it isn't dry when you start the motor. It's fine to do this, but if your filter is mounted at an angle it can make a mess. If your oil pump is functioning properly, the new filter will be filled within a few seconds, so pre-filling it isn't really necessary.

PROPERLY DISPOSING OF ENGINE OIL AND FILTERS

If you change your own oil, you must know how to properly dispose of the used oil and filter. Local regulations differ, so it's important that you know the rules in your area. In many places, you can put the used oil in your empty oil containers and take it to an auto parts store, where they will recycle the old oil for free. Some areas may require special recycling of used filters and oily towels. Check with your local dealer or auto parts store for more information on the local regulations.

How to: Inspect and Change Accessory Belts

The accessory belts are spun by the engine to drive features such as the alternator, power steering, and water pump. It is usually a serpentine belt that winds around several pulleys on the front of the engine. It will stretch over time, and can fail as the internal cords give way, or it cracks and breaks due to age. Belts with cracks, stretching, or fraying need to be replaced.

WHAT YOU NEED

- Replacement belt
- Tools to remove the belt and any accessories in your way (usually a socket wrench or socket)
- Safety gloves and glasses

A

B

The first sign of trouble with a belt is usually the noise. As a belt wears out, it loses tension and starts to slip against the pulleys it is driving, which creates a loud squealing noise.

INSPECTING THE BELTS

1. LOCATE THE BELTS.
   On most vehicles, the accessories are run opposite of the transaxle or transmission. This is usually on the left side or the front of the engine. Be aware that the belts may be obscured by a plastic cover (FIGURE A).

2. INSPECT THE BELTS WITH ENGINE RUNNING.
   With the engine running, watch the belt. Look for any kind of side-to-side wobble from the pulleys, or a noticeable wear point as the belt goes around (FIGURE B).

CAUTION

Keep hands and other items away from the running motor.

3. **INSPECT THE BELTS WITH ENGINE OFF.**

With the engine off, you can get a closer inspection of the belts. Put on your safety gloves and glasses and be careful when inspecting a warm or hot engine. The belts can wear on the inside and outside surfaces.

On the smooth side, look for evidence of slipping and grooves. If it looks glossy it may be slipping (FIGURE C). Deep grooves in the smooth side might denote a problem with one of the pulleys and a belt that is off center (FIGURE D).

On the grooved side, look for evidence of missing grooves or other damage. Any sign of reinforcement cords is a warning sign that the belt is failing. Try to inspect the belt when the grooved part is on the outside of a pulley. If the belt is starting to crack, the cracks will show up here (FIGURE E).

On both sides, look for internal belt cords and cracking. If you find any you will need to have the belt replaced (FIGURE F).

CHANGING THE BELTS

All cars are different, and how you change your accessory belt will depend in part on how it is built. Some cars have belts that are very easy to access, while others are much more difficult. This tutorial shows how to change a serpentine belt, or a single belt with V-grooves on one side. If your car has stretch belts, you will need a special tool to remove them.

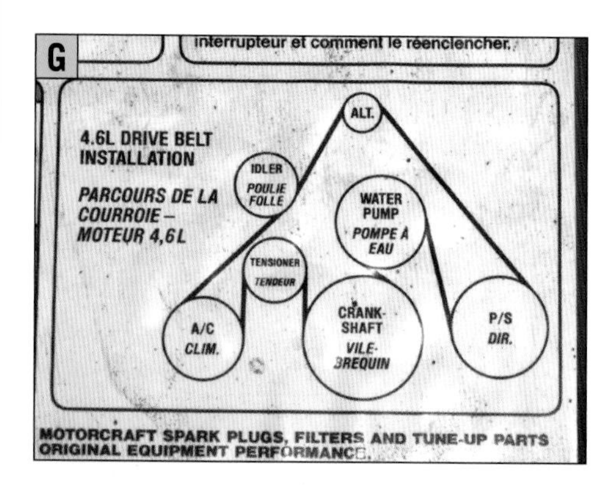

1. **LOCATE THE BELT DIAGRAM.**

   Most car makers put a sticker somewhere under the hood to show you how the belt is meant to be installed on the car. If your car does not have a label, make a diagram or take a picture so you will know how it should look when you replace the belt (FIGURE G).

2. **REMOVE THE BELT.**

   Most engines use a spring-loaded tensioner that keeps a belt tight as it wears and stretches. Some tensioners have an adjustment bolt that applies tension. If you pull the tensioner back, the belt will slacken and you can remove the belt. Some cars, like this one, use a square socket, so you can insert a socket wrench in the tensioner to twist it (FIGURE H).

   Pull the tensioner back to release the pressure on the belt and remove the belt. Once the belt slackens, you can pull it off with your hands. It may be a challenge to guide the belt around different parts to remove it from the engine.

TIP

When an engine throws a belt and it hasn't broken or stretched too far, it is possible that one of the pulley bearings is wearing out and the belt is sliding off of the pulley. Check your pulleys for proper straight alignment. Any kind of offset alignment will reduce the life of the belt and possibly cause it to come off the motor. Inspect the engine while it is running for any kind of wobble in the belt system.

STRETCH BELTS

Some manufacturers use stretch belts to eliminate the spring-loaded tensioner. A stretch belt works like a big rubber band, and requires a special tool to install and remove it. If your vehicle has stretch belts, use the proper tool for installation and removal. Using a screwdriver or other tool can damage the belt.

3. INSPECT THE PULLEYS.
   With the belt off of the engine, give the pulleys a spin to make sure they spin freely and don't wobble (FIGURE I). Wobble will cause the belts to come off of the engine. If you see wobble or if you feel grinding or grit as the pulleys turn, you will need to have the pulley or bearings replaced by a professional.

4. INSTALL THE NEW BELT.
   If your pulleys are okay, you can route the new belt using your diagram as a guide. Route all the paths up to the tensioner, then pull the tensioner back and install the belt, and then release the pressure on the tensioner. Inspect all the points where the belt meets each pulley to make sure it is properly aligned.

5. CHECK THE BELT.
   After you have installed the belt, run the engine and observe the belt as it moves. It should move smoothly over all the pulleys with no wobble.

THE DRIVETRAIN

IN THIS CHAPTER

How the Drivetrain Works

Common Drivetrain Problems

How to: Check and Fill an Automatic Transmission

How to: Check and Fill a Manual Transmission

How to: Check and Maintain CV Joints and U-Joints

How the Drivetrain Works

The drivetrain takes the rotating power from the engine or electric motor and transfers it to the wheels. It uses gears, shafts, and joints to transfer the rotation of the engine to the rotation of the wheel.

The primary component of the drivetrain is the transmission. Just like a multi-speed bicycle, the transmission changes the rotation speed of the drive axles with the speed of the engine. It allows more torque to start, and then reduces the torque once the car is moving at a higher speed. The following are some of the main components of the drivetrain.

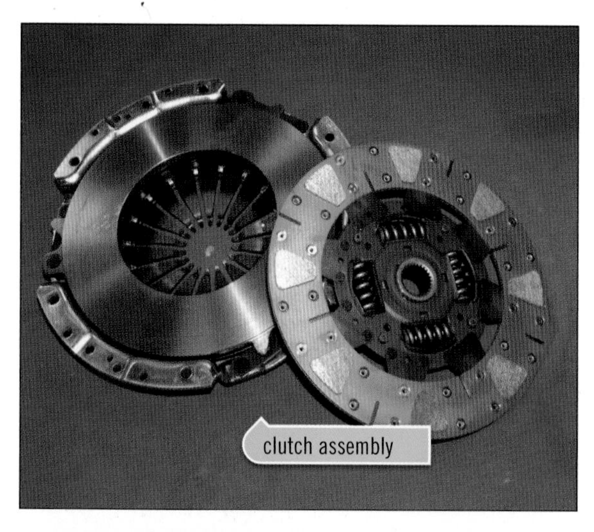

Torque Converter or Clutch A manual transmission uses a clutch assembly, and an automatic transmission uses a torque converter. Both do the same thing: they allow the transmission to be disengaged from the rotation of the engine and provide a connection between the engine and transmission.

Cars with a manual transmission use a clutch, which is a friction wheel and a spring loaded metal plate. When you push on the clutch pedal, you are pulling the spring loaded pressure plate away from the friction clutch disc, breaking the connection between the engine and the transmission.

A torque converter uses hydraulic pressure to engage the engine. As the engine speeds up, the torque converter creates more pressure and engages the transmission. The torque converter can also act like a gear reducer to increase torque at low speeds.

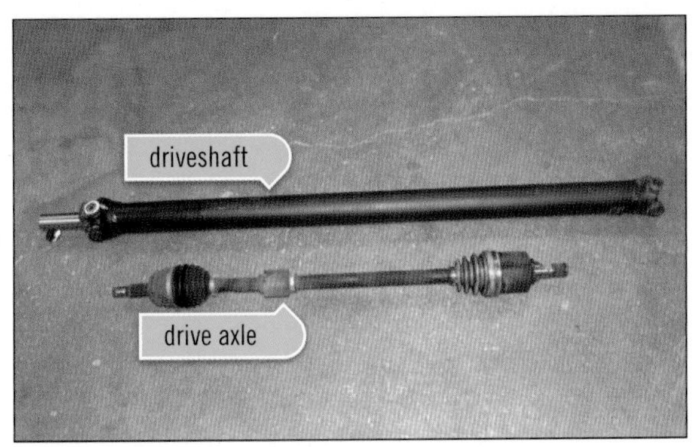

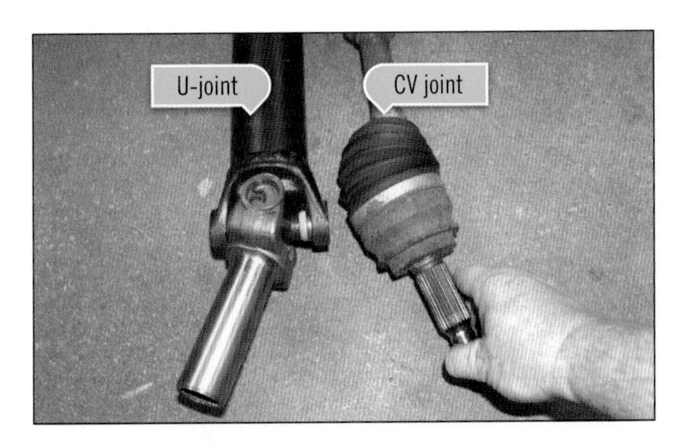

Transmission or Transaxle The transmission, also called a transaxle (front-wheel-drive vehicles) or a gearbox, uses different gear ratios to adapt the output of the engine to the speed of the wheels. Low gears turn the drive wheels slower. An overdrive gear turns the output of the transmission faster than the output of the engine. The transmission also allows the wheels to be spun in reverse.

A transmission sends its power to the drive axles with tubes called *driveshafts* and *axles*. A transaxle drives its wheels directly without any additional reductions.

Driveshafts and Drive Axles In rear- or four-wheel-drive cars, the transmission sends power to the axles through a driveshaft. In a front-wheel-drive car, the wheels are driven directly by the drive axles. Both use a type of flexible joint so the wheels can move without breaking the drive shafts.

U-Joints and CV Joints The use of universal joints, or U-joints and constant velocity joints, or CV joints, allow the drive shafts and axles to rotate at an angle and allow the drive points to move. U-joints are commonly used in rear-wheel and four-wheel drive cars, and CV joints are typically used in front-wheel-drive and cars with independent rear suspensions. CV joints typically are covered with a rubber boot to keep them clean.

Differentials and Axles On rear-wheel and four-wheel-drive cars, the axle uses an additional set of gears to convert the rotation of the driveshaft 90 degrees and out to the drive wheels. In a "solid" axle, the drive axles are housed inside the axle tubes. In an independent suspension system, the axles are usually exposed and use CV joints to allow each drive axle to move up and down.

Common Drivetrain Problems

Most problems with the drivetrain involve the transmission, which takes all that power and torque from the motor and turns it into rotational energy. An automatic transmission uses hydraulic pressure to shift itself, which adds complexity to the problem. Improper use of a manual transmission's clutch can lead to early failure, so if you have a manual transmission car, learn to shift properly. U-joints and CV joints last a long time but will wear out eventually.

LEAKS

The most common problem with the drivetrain is leaks. Most leaks occur because of the shafts going into and out of different parts of the drivetrain. Because these shafts are rotating, it takes a special type of seal to keep the fluids in, and they wear out from normal use. Most leaks show themselves as puddles under your car, and sometimes the leak may be two fluids mixed together, such as transmission fluid in the cooling system. Leaks left too long lead to failure, so if you find one, get it fixed as soon as possible.

DRIVETRAIN "SLIPPING"

When your engine is running just fine, but the car is not getting the power to the wheels, it can feel like the drivetrain is "slipping." This can be caused by a faulty automatic transmission, a worn-out clutch in a manual transmission, or the failure of the automatic torque convertor.

TRANSMISSION NOT SHIFTING GEARS

If you let an automatic transmission get too low on fluid, it won't be able to pump fluid through the transmission and torque converter to shift the gears. If an automatic stops shifting, you may hear the engine rev up higher than normal, or not feel the transmission shift down while decelerating

There are special gears in a manual transmission that let the gears slide in and out smoothly, called synchronizers. If the "synchros" or gears wear or become damaged, you may not be able to put the transmission into the correct gear. If the clutch is worn, you may not be able to disengage the transmission from the engine to shift gears.

The fluid lubricates all the gears in both manual and automatic transmissions, and if it gets too low the transmission can destroy itself.

TRANSMISSION POPS OUT OF GEAR

A transmission that pops out of gear can be dangerous, not just inconvenient. If the synchros wear out, the transmission may slip out of gear. You'll know this when the engine revs up and the car starts coasting.

OVERHEATING

An automatic transmission or transaxle uses a cooler to cool the hydraulic pumps, and if it overheats, the transmission can fail. When looking for leaks, make sure you look around the radiator for transmission leaks. A transmission that is overheating may emit a burning smell.

U-JOINT AND CV JOINT FAILURE

U-joints and CV joints need to be lubricated regularly since they use roller bearings to provide rotation. When a U-joint or CV joint is failing, it will start vibrating. You'll notice this vibration while driving at speeds between 40 and 55 mph (65 to 88 kph).

The U-joints and CV joints are transferring power from one shaft to another, and if they get sloppy they will start to "clunk" as power is applied to them and occasionally when the transmission is changing gears.

If the boots that protect the CV joints boots tear, dirt can get in and grease can get out. A torn CV joint boot needs to be replaced or the joint will fail prematurely.

GRINDING OR OTHER NOISES

You may hear the sound of metal grinding against metal if the fluids in the transmission or differential are low, leaking, or worn out. A grinding sound means that metal is being torn apart, and you need to have it checked immediately.

"CHECK ENGINE" LIGHT COMES ON

Your computer may sense if the transmission is not functioning properly, causing the "check engine" light to come on.

How to: Check and Fill an Automatic Transmission

An automatic transmission, or transaxle, uses hydraulic pressure to change gears up and down. The transmission uses transmission fluid and a pump to make the pressure and move the parts.

WHAT YOU NEED

- Paper towels
- Transmission fluid (if needed)
- Funnel (optional)

Since about 2007, some car makers have stopped installing a means to check the level of fluid in the transmission. Those transmissions are sealed, and some do not have a filter. Instead, they rely on a sensor to tell you when the fluid is low, and require a complete fluid change at regular intervals.

If your car does have dipstick tube and a means to check the fluid, here's how you do it.

CHECKING THE FLUID LEVEL

1. LOCATE THE TRANSMISSION DIPSTICK.
   On this front-wheel-drive car it is well-marked and easy to access. It may be harder to find on a rear-wheel-drive car. Check your owner's manual for the location (FIGURE A).

2. TURN ON THE VEHICLE AND ALLOW IT TO RUN.
   This allows the pumps to push the fluid into all the passages in the transmission. Be careful to not get near any rotating parts on the engine.

3. CLEAN THE DIPSTICK.
   With the car running, pull the dipstick, wipe it off, and reinsert it. This will give you a fresh reading while the transmission is at normal operating levels.

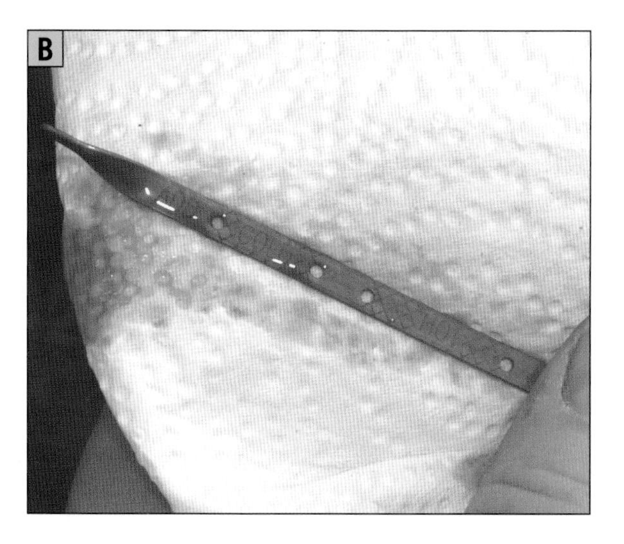
B

4. PULL THE DIPSTICK AND CHECK THE LEVEL.
   This dipstick has readings for when the transmission is cold or if it is hot. The fluid in the transmission will expand when warmed up, so check your owner's manual and see if it needs to be checked when warm.

   Examine the condition of the transmission fluid. If it is black with burnt material or soot, or if it looks milky or hazy, you might need to have it changed or have the transmission serviced (FIGURE B).

   If the level is okay, replace the dipstick and turn off the car.

FILLING TRANSMISSION

If your transmission fluid is low, you can add more. Take care not to overfill the transmission, and be sure to use the fluid recommended by the manufacturer.

C

1. DETERMINE WHAT TYPE OF FLUID YOU NEED.
   Some manufacturers print the type of fluid you should use on the dipstick or tube (FIGURE C). This will also be in the owner's manual.

D

2. FILL THE TRANSMISSION.
   With the engine turned off, insert a funnel in the dipstick tube and pour in the fluid. A long funnel is good for tubes at the back of the engine that are hard to get to. Go slow. It's better to add a little, recheck, and then add more than to put in too much (FIGURE D).

3. RECHECK THE LEVEL
   Let the fluid run down the tube for a couple of minutes, and then recheck the level of the transmission. Continue until you have filled the transmission to the required level.

How to: Check and Fill a Manual Transmission

Manual transmissions, transaxles, and differentials are big gear sets that don't have pumps like automatic transmissions. They lubricate the gears by dipping down in the oil and slinging it all over the inside of the housing. Manual gears don't usually require external cooling.

WHAT YOU NEED

- Paper towels
- Gear oil or transmission fluid
- Wrench or hex wrench
- Oil drain pan
- Jack and safety stands
- Tubing or fluid pump (optional)
- Thread sealant (optional)

Filling a manual gearbox is easy to do, but sometimes hard to access. Make sure you're able to locate and access your transmission before you begin. Be sure to recycle any used fluids properly.

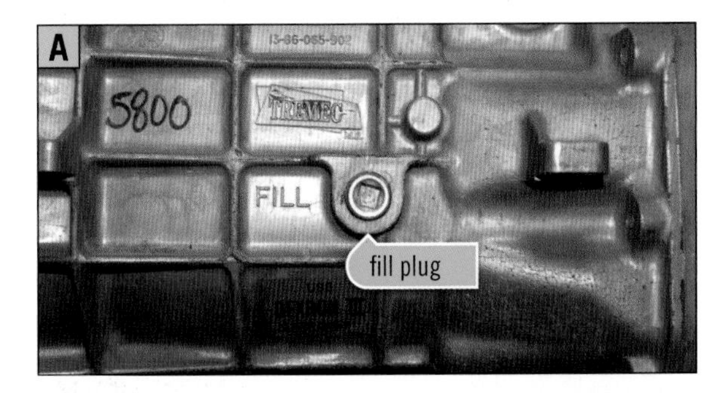

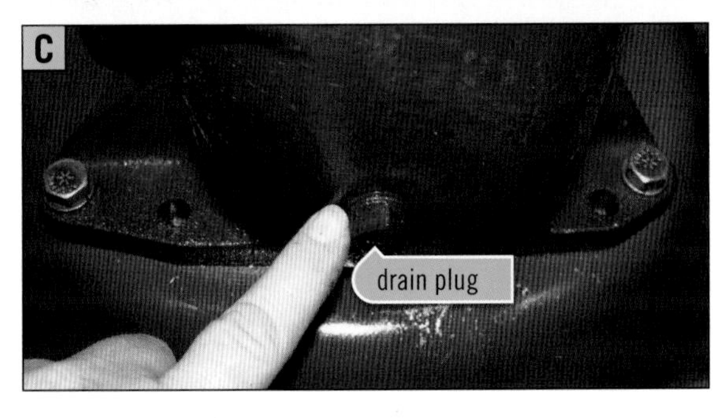

1. RAISE AND SECURE THE CAR SAFELY.
   Use a jack and safety stands to raise the car. Make sure the car is secure.

2. LOCATE THE FILL AND DRAIN PLUGS.
   The fill plug may be located on the side or back of the transmission or differential and may be identified with the word "Fill" (FIGURES A AND B).

   The drain plug is usually located on the bottom of the housing (FIGURE C).

TIP

Some differentials do not have a drain plug, but are emptied when the gear set is removed from the differential, which usually requires pulling the axles. If this is the case with your car, better it's better to leave this repair to a professional.

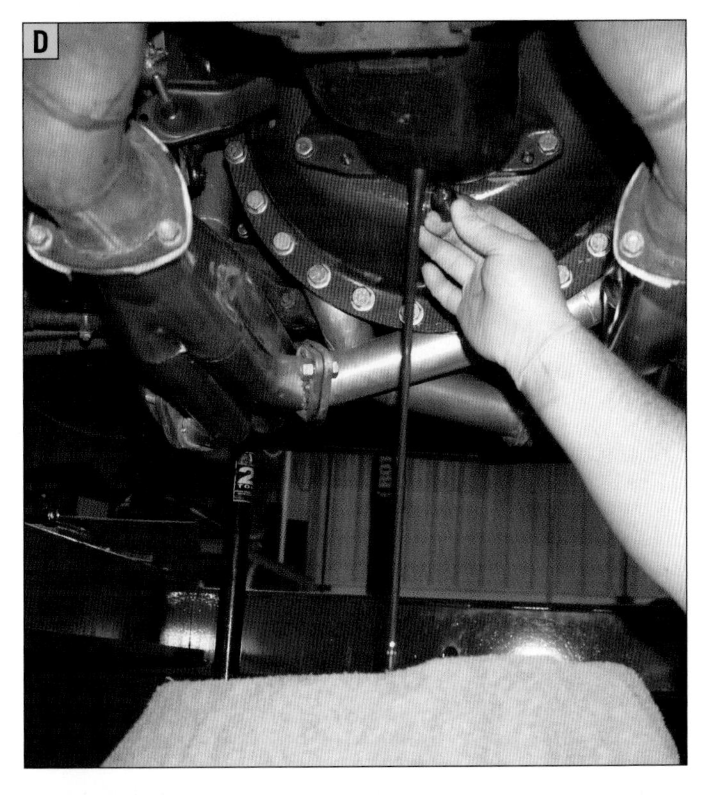

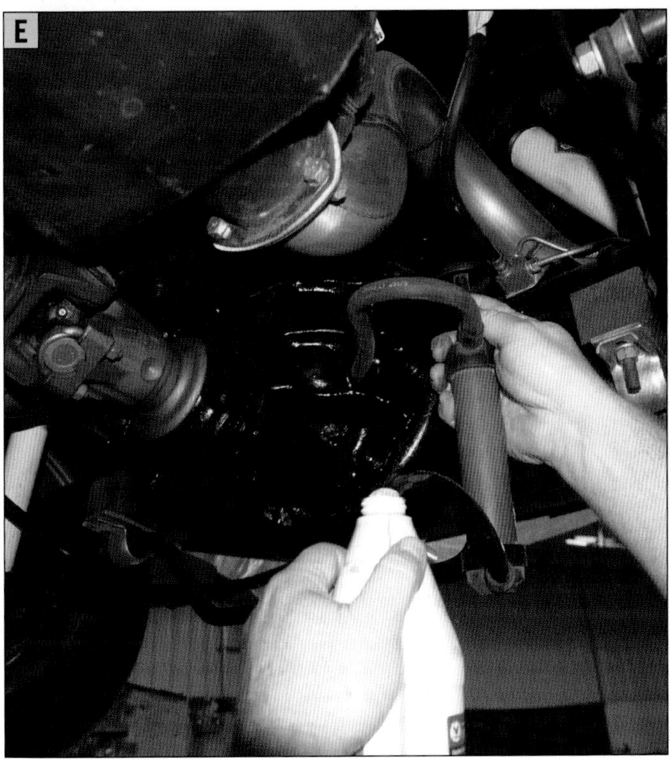

3. DRAIN THE FLUID.

It is not necessary to drain the fluid in the transmission unless it needs to be changed per the manufacturer's recommendations. To do so, remove the plug with your wrench and allow the fluid to drain completely from the housing into an oil pan or bucket (FIGURE D).

Place a filter, a piece of cheesecloth, or a clean rag between the pan and housing to collect any loose metal shavings or contamination. If you see a lot of metal shavings, have a professional inspect the transmission. After draining, reinstall the drain plug.

4. FILL THE HOUSING.

If the old fluid came out clean, remove the fill plug and begin filling the housing with new fluid. Getting a bottle into confined areas can be difficult. You can install a piece of hose to the end of your fill bottle or use an inexpensive fluid pump to help push the fluid into the housing (FIGURE E).

5. APPLY SEALANT.

Once the fluid starts running out of the fill tube, apply a little thread sealant to the plug and re-install the plug tightly.

How to: Check and Maintain CV Joints and U-Joints

CV joints and U-joints allow the wheels driving the car to move up and down with the suspension. CV joints, which allow for greater movement, are generally used on front-wheel-drive cars and cars with independent rear suspensions. U-joints are generally used on rear-wheel-drive vehicles and larger four-wheel-drive trucks. Both allow the joint to bend while the drive tubes rotate.

DRIVING TEST

The first step in inspecting the drive joints is to drive the car and listen and feel for problems. Look and listen for the following when driving the car:

1. VIBRATION WHILE DRIVING.
   A failing joint may cause vibration at specific speeds. Some U-joints will be smooth at low and high speeds, but start to vibrate around 40 to 55 mph (65 to 80 kph). The vibration may also increase with speed. If you are experiencing a vibration, stop the car and shift into neutral or park and rev the motor to make sure the vibration is not engine related.

2. A CLUNKING SOUND WHEN GOING INTO OR CHANGING GEARS.
   If you feel a clunk when the car starts from a stop or when the transmission changes gears, the joints are loosening and wearing, causing the torque of the transmission to slap the loose joint.

3. A POPPING NOISE.
   A constant click or pop coming from one of the joints means part of the joint has worn and is catching on each revolution.

4. A GRINDING OR METAL-ON-METAL SOUND.
   A joint where the grease has failed will begin to disintegrate and produce a metal-on-metal sound.

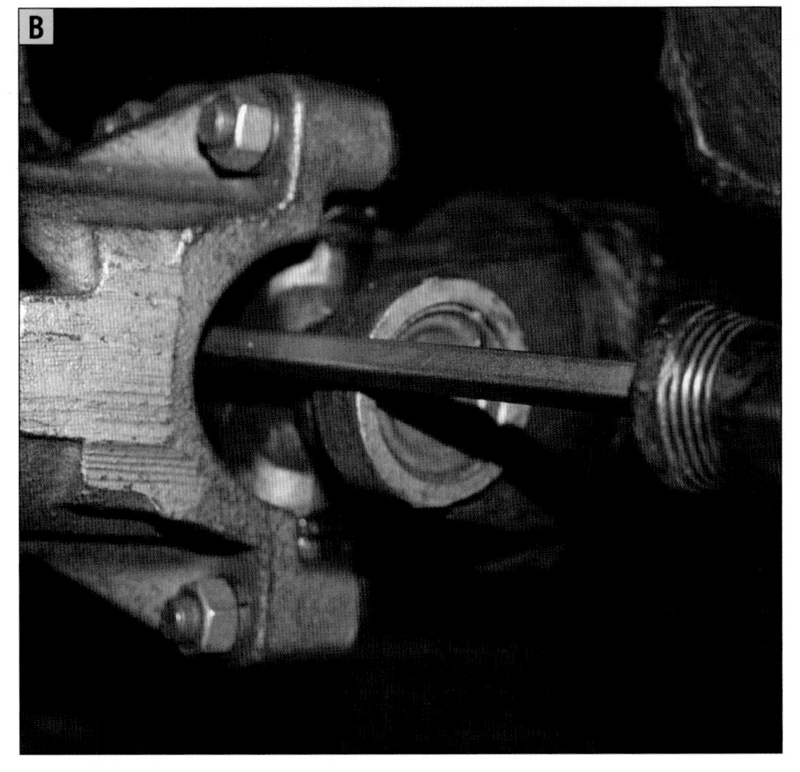

INSPECTING U-JOINTS

There should be no play in a U-joint. You can inspect the joints by trying to turn the driveshaft by hand to see if it moves.

1. **RAISE THE VEHICLE**
   If necessary, raise the vehicle and secure with jack stands in order to access the driveshaft.

2. **WIGGLE THE DRIVESHAFT.**
   With the vehicle in park, try to rotate the shaft. The U-joints should have no play and the tubes should all rotate together. You may see some movement in all the components. Wiggle the driveshaft up and down and side to side to look for movement. Test both ends of the shaft (FIGURE A).

3. **TEST THE END PLAY.**
   You can check the end U-joint by gently applying leverage between the joint and the differential with a screwdriver. It should remain firm and not move at all (FIGURE B).

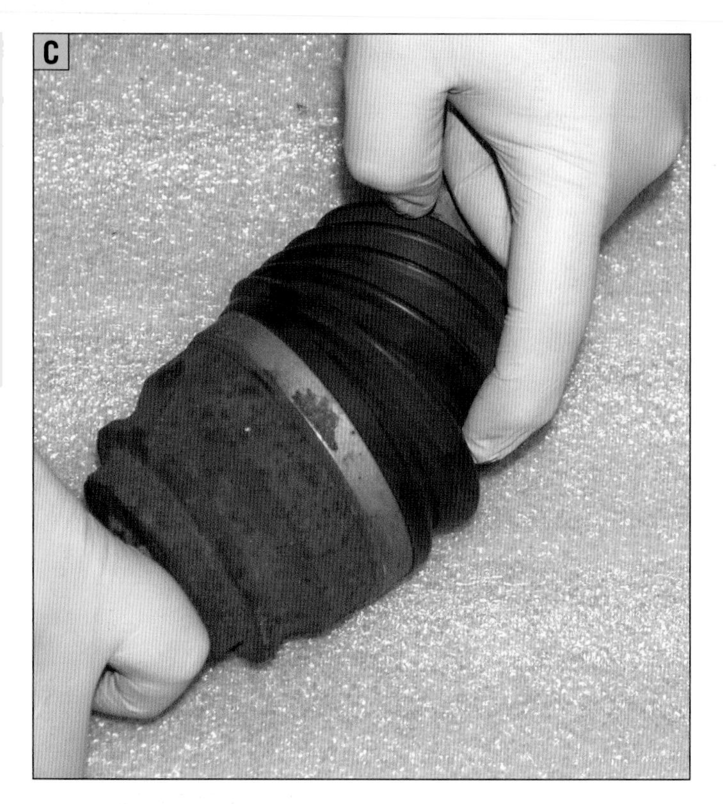
C

D

INSPECTING CV JOINTS

CV joints are sealed in a protective rubber cover, or boot, and if the boot becomes torn or breaks, the grease inside will leak out. A torn CV boot can be replaced, but it requires some knowledge to do so.

1. RAISE THE VEHICLE.
   If you can, raise the vehicle and secure with safety stands so the drive wheel being tested is off the ground. This will allow you to turn the wheels fully to inspect the boots.

2. INSPECT THE BOOT.
   Put on protective gloves and run your finger around the boot, feeling between the grooves for grease (FIGURE C). Globs of grease indicate that the boot may be torn and is forcing grease out of the tear (FIGURE D).

3. ROTATE THE DRIVE AXLE.
   With the wheel turned out, put the car in neutral and slowly spin the drive axle to view the entire boot in the open position. The side that is compressed can hide a tear.

   As you rotate the shaft, listen for grinding or popping noises.

4. WIGGLE THE SHAFT.
   As with the U-joint, wiggle the output shaft and look for joint movement. It should be tight.

GREASING A U-JOINT

As part of your normal maintenance, it is good to keep the U-joints greased if they are equipped with a grease fitting. Some heavy duty U-joints are not equipped with grease fitting because they can crack and fail at the fitting. The U-joint is designed to extrude the used grease at the ends, purging the used grease from the joint.

WHAT YOU NEED

- Grease gun and grease
- Paper towels or a clean rag
- Gloves and eye protection

1. **LOCATE THE GREASE FITTING.**
   Raise and safely secure the vehicle if necessary. The grease fitting is a small, ball-shaped nipple with a valve in the end to keep the grease in (FIGURE E). Use a paper towel or clean rag to wipe dirt and grime from the fitting. Also, clean all dirt off the grease gun so no contaminants get pushed into the fitting.

2. **ATTACH THE GUN TO THE FITTING.**
   You should feel the end of the grease gun snap over the fitting (FIGURE F).

3. **GREASE THE U-JOINT.**
   Begin pumping fresh grease into the U-joint until you see grease coming out of all four seals. Because this U-joint isn't in a car, the grease came out of the two cups that aren't secured (FIGURE G). It should come out of all four seals simultaneously. Once you see fresh grease coming out, remove the grease gun and wipe away the purged grease and the grease around the fittings.

THE FUEL SYSTEM

IN THIS CHAPTER

How the Fuel System Works

An engine requires three things to create energy: fuel, air, and an ignition source. The engine draws in air and then adds fuel, which is mixed with the air and burns in the engine cylinder to turn the engine. The fuel system is responsible for making sure the engine has the right mixture of air and fuel to burn.

The fuel injection system of a modern engine is very different from what was used 30 years ago. In the old days, the engine used the moving piston to create a vacuum, pulling fuel out of a carburetor and mixing it with atmospheric air. Today, computers can sense when fuel is needed, and spray it directly into the engine cylinder. Although pistons are still used to draw air into the cylinder, today's engines run much more efficiently.

THE FUEL SYSTEM

Fuel Injector While fuel injection has been around for more than 100 years, it wasn't until computer controls came along in the 1980s that fuel injection became the preferred choice for better performance and efficiency.

A fuel injector sprays fuel into the engine, rather than having it drawn out from an engine vacuum like a carburetor. The computer can control how long and when to trigger the fuel injector.

Terms to know:

- **Central fuel injection** One or more fuel injectors are centrally mounted on the engine and the fuel is dispersed to all the cylinders evenly.
- **Multi-port fuel injection** There is a single injector for each cylinder and the computer can turn the injectors on and off, usually in batches.
- **Sequential fuel injection** A multi-port setup, but the computer can select exactly when to fire the fuel at the right moment to each cylinder for maximum power and efficiency. Most modern cars have this feature.

Fuel Pump The fuel pump moves the fuel from the tank to the engine. It can be mounted in the fuel tank, just outside the fuel tank, or even on the engine. Some fuel pumps run continuously, and if the fuel pressure gets too high, a valve opens and the extra fuel is sent back to the fuel tank. Some fuel pumps are controlled by the computer and can be turned on and off, as needed.

Fuel Tank The gas tank does more than just hold the fuel; it's part of the emissions system of your car. In the old days, the gas vapors in the gas tank were vented out into the atmosphere. Today, there is a valve in the gas tank that allows these vapors to be funneled back to the engine and burned as part of the combustion process. This system is called the evaporative emission control (EVAP) system.

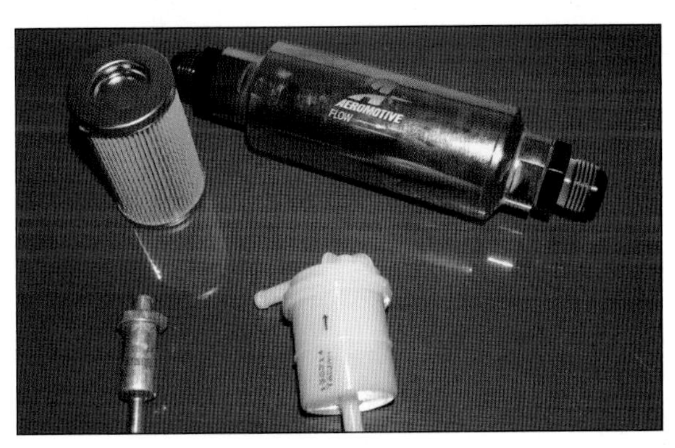

Fuel Filter Fuel filters come in all shapes, sizes, and varieties. Some are self-contained, some are cartridges, some attach to the engine, and some to the tank. Where yours is located and how often you need to change it depends on your manufacturer. The fuel filter pulls out small particles in the fuel, preventing them from clogging the fuel injectors and pump, and keeps contaminants out of the engine.

Fuel Inertia Switch This switch cuts off power to the fuel pump if it is jostled in an accident. Since the fuel pump is electric, it may still pump fuel after a wreck, and this can cause a dangerous fire hazard, so this switch is used to shut off the fuel flow. It can be reset by pressing the button on the top of the switch.

Fuel Sensors and Monitors Your car relies on a variety of sensors and monitors to determine when to deliver fuel to the engine. These include:

- **Fuel rail transducer** The computer uses this feature to measure the pressure in the fuel lines, and adjusts the power to the fuel pump when needed.
- **Fuel pressure regulator** A mechanical way to regulate the proper fuel pressure. If the pressure gets too high, the regulator will open up and send the excess fuel back to the fuel tank. This is used when the pump is not controlled by the computer.
- **Fuel tank pressure sensor** This tells the computer if the pressure in your gas tank is correct. If it's too high or too low, it means the vapors probably aren't getting to the engine to be burned.

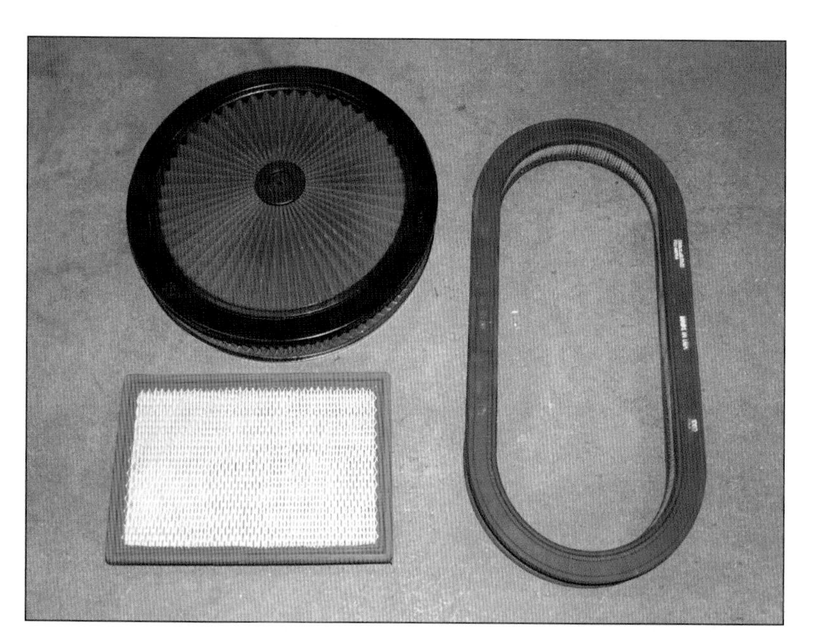

AIR INTAKE SYSTEM

Air Filter This is your first defense against debris getting pulled into your engine. Most filters are disposable, although some filters can be cleaned and reused. The shape, size, and design of air filters varies greatly by manufacturer.

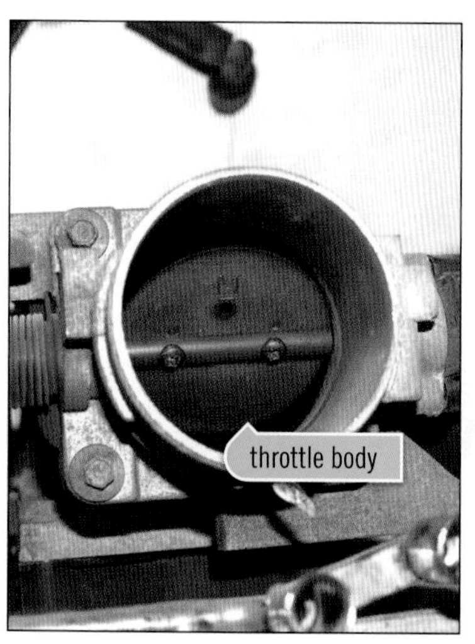

Throttle Body The throttle body opens when you step on the accelerator pedal. It has a big plate inside a tube that allows air into your engine. As you press on the pedal, the throttle plate moves into a horizontal position, and air is allowed into your engine.

Throttle Position Sensor This tells your computer how open the throttle plate is and how much air is getting into the engine.

EGR Valve The full name for this component is the *exhaust gas recirculator valve*. Just as your engine burns fuel tank vapors, it also takes a portion of the exhaust gases and sends them back into the engine to be burned again. This helps reduce engine emissions.

PCV Valve PCV stands for *positive crankcase ventilation*. This one-way valve is another device designed to reduce emissions. As fuel and oil vapors build up in the engine, they build pressure in the engine crankcase. The PCV valve opens up to allow these vapors out of the engine crankcase and injects them into the intake manifold to be burned, relieving pressure buildup.

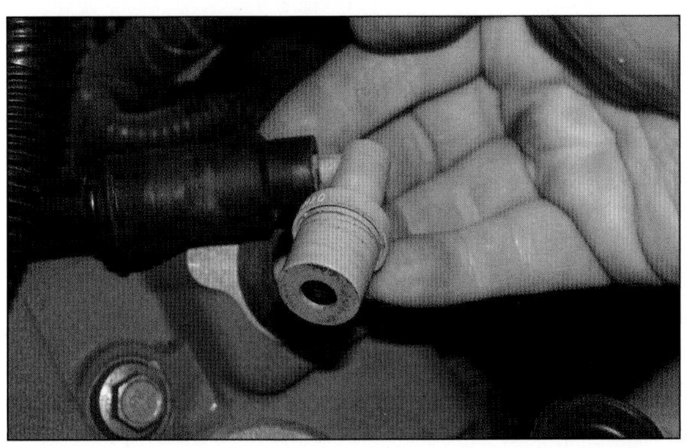

Intake System The intake system consists of all the plumbing that comes before and after the throttle body. This includes the tubing between the air filter and the throttle body (on the right) and the manifold that disperses the air to the individual cylinders (on the left). Intake manifolds usually have individual pipes, called *runners*, that lead to each cylinder. This helps the air move faster and allows more air to enter the cylinder.

Mass Air Flow (MAF) Sensor This sensor tells the computer how much air is going into the engine. Air density can change due to altitude, climate, and whether your car uses a supercharger or other means to force air into the engine.

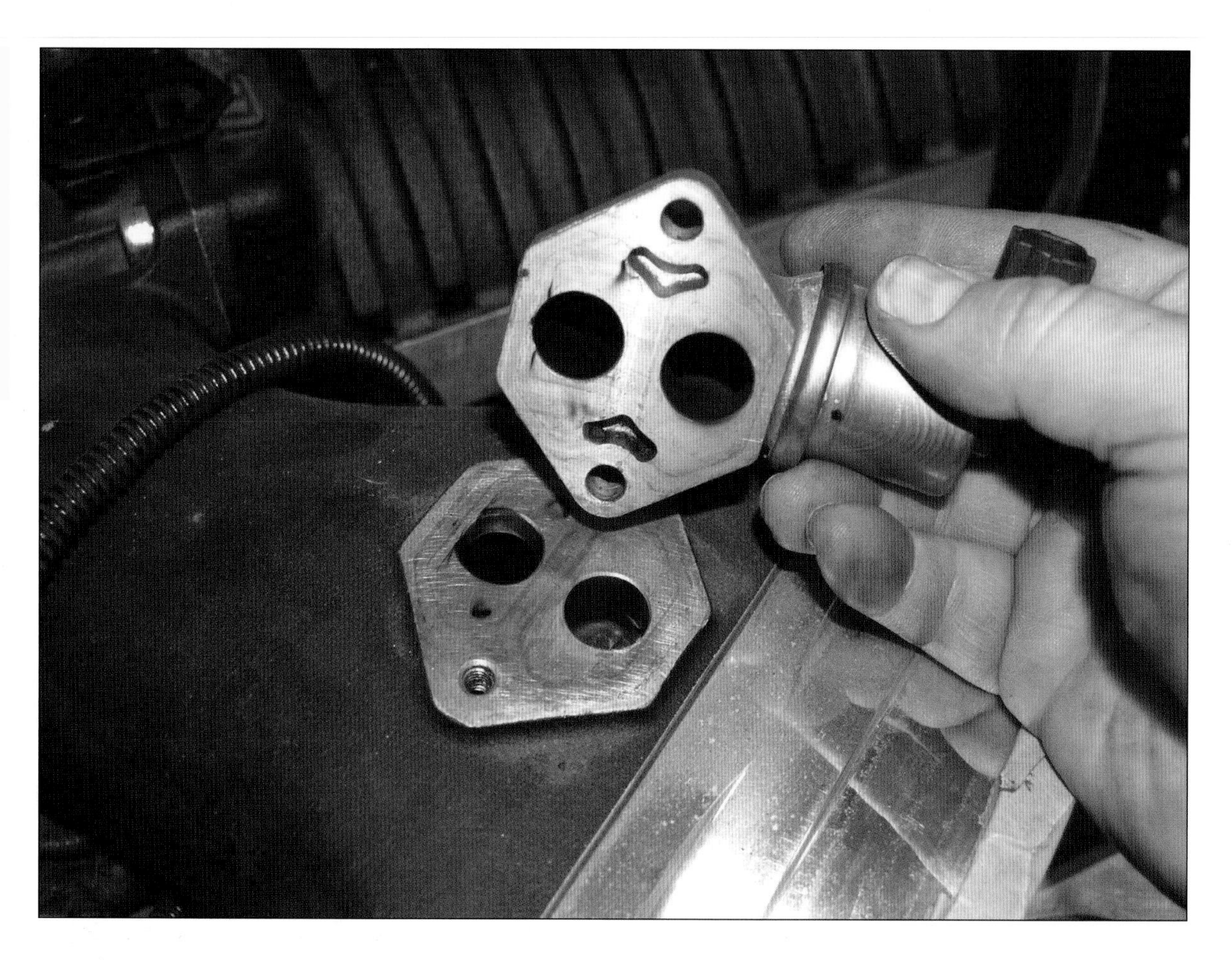

Idle Air Control (IAC) Valve When the throttle plate is closed, this valve allows a small amount of air to bypass around the throttle plate so the engine doesn't get choked off.

Air Sensors and Monitors Your computer uses information from several sensors to determine how much fuel to add to the engine. Your mechanic may refer to the following sensors when talking about your intake system.

- **Barometric Air Pressure (BAP) sensor** This measures the density of air, like a MAF sensor, but measures just barometric pressure. At sea level, air is denser than at higher altitudes, and this sensor is used by some systems to tell the computer how dense the air is that's entering the engine.
- **Intake Air Temperature (IAT) sensor** This measures the temperature of the air. Colder air will be denser than hotter air, and cold air requires more fuel, while hotter air needs less fuel but can cause the fuel to burn too soon (called *pre-detonation*).
- **Electronic Throttle Control (ETC) motor** On cars that have an electronically controlled accelerator pedal (throttle by wire), this motor is used to move the throttle plate in place of a cable.

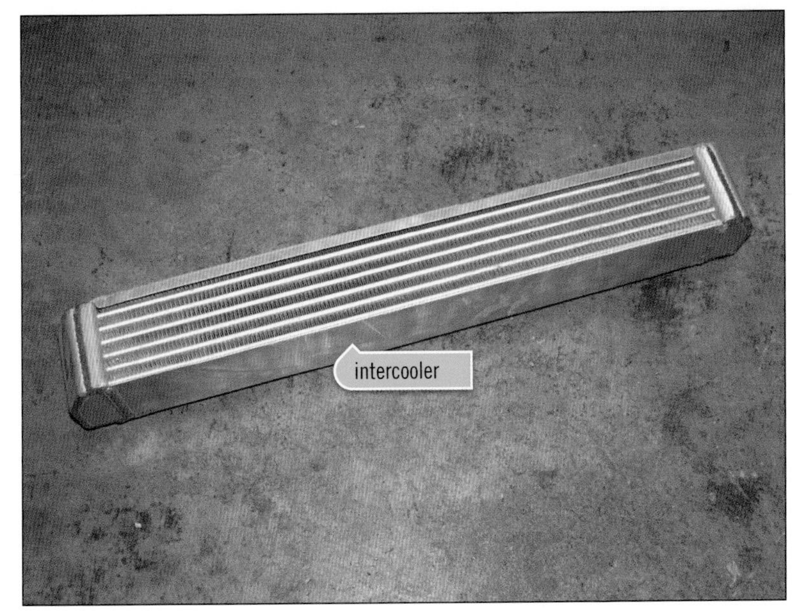

SUPERCHARGERS, TURBOCHARGERS, AND INTERCOOLERS

Superchargers and turbochargers are big air compressors, and they force large amounts of air into the engine and to create a denser fuel and air mix. This increases the power of the engine.

A supercharger is driven by the engine, usually by a belt on the front of the motor. Turbochargers are driven by the exhaust gases from the engine turning an impeller (like a pump) that is connected to the compressor. Turbochargers have something called "turbo lag" which means they have to wait until the gases are running fast to get the compressor to spin, while the supercharger can start building pressure almost immediately. However, a supercharger also draws power away from the engine due to the additional load on the engine, while using the exhaust gases doesn't affect the load on the engine.

When you compress air, it heats up, and hot air can lead to the pre-detonation. Intercoolers are used to cool the air before it is drawn into the engine. An intercooler is a radiator, and the intake air is drawn through it and cooled, just like the coolant in the engine radiator.

Common Fuel System Problems

Fuel and air problems are easy to describe, but difficult to diagnose. If the engine is not getting enough of one or both, it will start to run poorly and may vibrate or have reduced power.

LOW OR NO AIR DELIVERY

A modern engine really needs to know exactly how much air is drawn into the engine, how dense it is, and what temperature it is to allow for efficient combustion.

Clogged Air Filter A clogged air filter means the engine can't get enough air, which causes it to reduce the amount of fuel delivered and can result in the engine losing power. Check your filter on a regular basis.

Dirty MAF Sensor The MAF contains a small wire that heats up to measure the density of the air going into the engine. If the wire gets dirty, it will not read properly. The MAF can be cleaned, but you need to buy a special cleaner from the auto parts store that's made specifically for cleaning the sensor.

Faulty Throttle Position Sensor The throttle body is a mechanical plate, so it usually doesn't have problems, but the sensor that tells the computer how open it can fail. The TPS may have to be calibrated and aligned; if so, leave this replacement to a professional.

Faulty EGR System The EGR valve can become clogged and fail. The engine may run rough, stall, or not run at all.

Clogged or Sticking PCV Valve Like the EGR, the PCV valve can become clogged or stuck. If this happens, the engine will start to lose power or not accelerate, or the pressure can build up in the crankcase and cause seals to fail. Checking the PCV is easy and quick, and in most cases relatively inexpensive to replace.

Clogged or Failed IAC The IAC allows a small amount of air to bypass the throttle plate at idle. If the valve fails or clogs, the engine may come off acceleration slowly, or it may stall and come down too fast. If it is closed, the car may not start and run at idle.

LOW OR NO FUEL PRESSURE

When the engine starts losing fuel pressure, the injectors can't deliver enough fuel to allow the engine to run smoothly. Many times the engine will start vibrating if it starts to lose power.

Loss of fuel pressure can happen at any point in the fuel system—the screen inside the fuel tank can be clogged, the fuel pump itself can fail, the fuel filter can be clogged, the lines can be blocked, or the fuel injectors can be dirty.

Clogged Fuel Filter If your car has a replaceable filter, it can be checked and replaced relatively inexpensively. Some newer cars do not have a replaceable fuel filter, as it is part of the fuel pump assembly mounted within the gas tank of the car.

Faulty Fuel Pump Some fuel pumps use plastic gears, and running the incorrect fuel can destroy the pump by dissolving the plastic parts. This can happen when you try to run alcohol-based fuels in a car not designed for them.

The fuel pump may also stop running because it wears out, or the electricity running the pump may fail. There are usually a fuse and a relay powering the fuel pump, and these go bad more often than the pump does.

Clogged Fuel Injector The fuel injector has a little needle opening that moves back and forth. Small particles of dirt occasionally get past the filter and may build up on the injector, or poor fuel can gum it up and prevent it from working. Contaminants can also partially block the opening and prevent the fuel from spraying out properly (atomizing), resulting in the fuel not mixing with the air as it should.

One preventative maintenance step you can take is to run a bottle of fuel injection cleaner through your engine every time you have the oil changed. However, if you have a clogged injector, you will have to use a professional-grade cleaner, or have it cleaned by a professional.

Open Gas Cap One of the most common computer codes comes from leaving the gas cap off of the gas port, which means the vapors from the tank are not being drawn into the engine through the EVAP system.

Inertia Fuel Switch Some car makers use a switch to shut off power in a wreck. If your car gets jostled, the switch can trip and close. If the car won't start, try resetting the switch. Some car makers use oil pressure or other means to shut off the fuel.

Clogs and Leaks Inspect your fuel lines for leaks on a regular basis. Fuel lines are made out of solid and flexible materials, and both can fail. Flexible lines wear out and hard lines can be kinked or rusted.

Sensors Since your car relies on sensors to know how much fuel is being delivered, the car won't run properly if a sensor goes bad. When a sensor goes bad, the computer may send an error code that the sensor is out of range.

How to: Change the Fuel Filter

Not all cars have a replaceable fuel filter. For those that do, the difficulty in changing the filter depends on where it is located. It may be mounted in the engine compartment, or underneath the car along the fuel path to the engine. Before buying a replacement filter, check your shop manual for the exact location and tools needed to change the filter.

WHAT YOU NEED

- Replacement fuel filter
- Gloves and safety glasses
- Jack and jack stands (if needed)
- Tool(s) to remove the fuel lines from the filter
- Tools to remove the mounted filter from the car
- Pan or tray to catch fuel from the car
- Container for recycling used fuel
- Paper towels or clean rags (optional)
- Fuse puller (optional)

1. **REMOVE THE FUEL PUMP FUSE.**
   Pulling the fuel pump fuse will release the fuel pressure and prevent fuel from spraying on you or the car. To do this, pull the fuel pump fuse or relay located in the fuse box. Check your owner's manual for the correct fuse or relay (FIGURE A).

2. **START THE ENGINE AND LET IT DIE OUT.**
   This should only take a second or two. Turn the engine off and leave the fuse out while you work on the fuel filter to prevent accidental spraying of fuel.

CAUTION

Fuel for your engine is highly explosive and the vapors can ignite very easily. Fuels can cause chemical burns if you are exposed to them. Take great care while handling fuels, and learn the emergency procedures before you come into contact with them.

TOOLS TO REMOVE THE FILTER

Some filters require the use of a special removal tool. Find out what kind of tool your filter needs before starting the job.

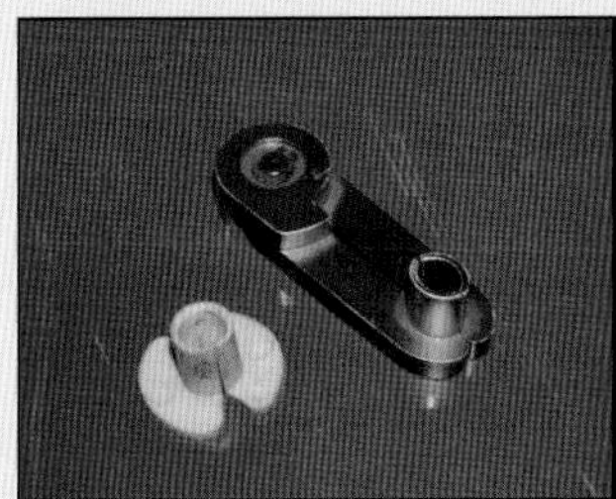

These tools press against a spring that's mounted inside the filter or the fuel line.

This filter has two different types of connectors to fasten the it to the fuel lines. The connector on the right requires a fuel line tool, and the two on the left are spring-loaded.

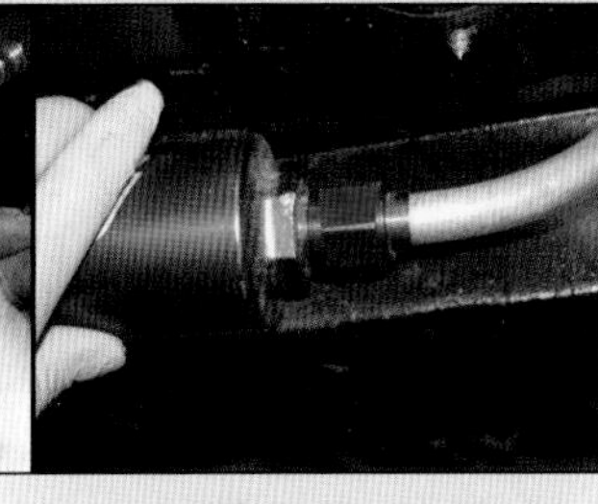

This racing filter uses a wrench to remove the lines.

3. LOCATE THE FUEL FILTER.
   The fuel filter can be located anywhere along the path of the fuel lines. If it is located under the car, raise the car and place it on jack stands. This fuel filter is mounted to the underside of the vehicle, close to the gas tank (FIGURE B).

4. REMOVE THE FUEL LINES FROM THE FILTER.
   Place a pan under the filter to catch any fuel and remove the lines from the filter. This filter uses a pair of plastic clips that hold the lines in. They are one-time use, and since the new filter came with replacement clips, we used a screwdriver to pull them out (FIGURE C). Take a picture to note which way the clips are oriented.

5. REMOVE THE FILTER.
Some filters are bolted down, some are simply held by the fuel lines. This one is held in place with a band clamp. Loosen the mounting hardware and remove the filter. Be careful as you pull the filter; do not splash fuel (FIGURE D).

6. DRAIN AND INSPECT THE FILTER.
Drain any remaining fuel out of the filter and inspect what comes out of the inlet of the filter. This filter was not clogged, but it was time for it to be changed. The contaminated fuel coming out of it means the filter did its job (FIGURE E).

FUEL PRESSURE AND RELEASE VALVES

Newer cars use the computer and a sensor to tell the mechanic if the car has proper fuel pressure. In the old days, cars had a Schrader valve mounted on the fuel rail. By hooking up a fuel pressure gauge to the Schrader valve, you could check the pressure at the engine.

If you do want to check the pressure on your older vehicle, this is where you hook up. When you start your motor, the fuel pressure should go slightly higher than the normal operating pressure, then settle back down once the engine is up and running.

BE CAREFUL—this port is usually on the top of the engine and can spray hot fuel and cause an explosion or a fire. Don't use this port to release the fuel pressure; just pull the fuse.

7. INSTALL THE NEW FILTER.
Most fuel filters are directional and should only be installed one way. This one is marked with an arrow to show the direction of fuel flow to the engine (FIGURE F).

Installation is the reverse of removal. Make sure the fuel lines are tight and the clips snap in place (FIGURE G). Once the fuel lines are connected, you can remove the car from the safety stands, if needed.

8. CHECK FOR LEAKS.
Reinstall your fuse then turn the key to the "on" position, but do not try to start the motor. The fuel pump will start and run for a second or two to reestablish pressure. Check your filter connections for leaks. If none are found, start the car and let the air purge from the fuel system. If you have a leak, re-check your connections for proper fit (especially if you have screw-on style filter connections).

9. DISPOSE OF THE OLD GASOLINE.
Check your local recycling regulations on disposing of the gas and filter from the filter change.

CAUTION

Don't put the dirty stuff back in your tank. Gas vapors are highly explosive, so take great care when transporting gasoline, and use approved containers to move the fuel.

How to: Inspect and Replace the Air Filter

The air filter is your first line of defense to prevent dirt and contaminants from entering the engine. Replace it regularly, according to the recommendation in your owner's manual. If you frequently drive on dusty dirt roads, you should change your filter more often than the recommended schedule.

WHAT YOU NEED

- Tools to open the filter air box (if needed)
- Replacement filter

A

B

1. LOCATE THE AIR FILTER BOX. With the engine cold, locate the filter box. It may be mounted under additional covers. The cover will usually stay connected to the air intake tubing. Depending on the manufacturer, you may have to remove the air intake tube from the box (FIGURE A).

2. INSPECT THE FILTER. Check the top side of the filter for dirt and contamination. Inspect the seal surrounding the filter for damage. If the seal has slipped or torn, it will allow dirty air to bypass the filter (FIGURE B).

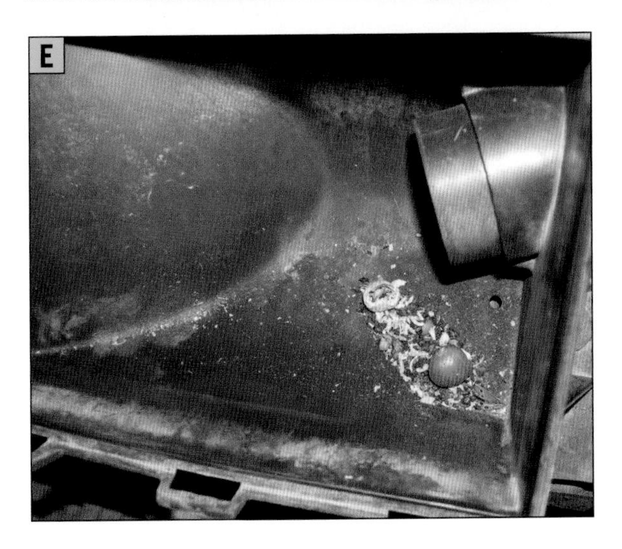

3. REMOVE AND INSPECT THE FILTER.
   You should be able to pull the filter out by hand. Inspect the filter in between the pleats for tears and damage (FIGURE C). Big pieces of contamination can be brushed out with a very soft bristle brush.

4. LIGHT CHECK THE FILTER.
   Hold the filter up to a strong light to see if the filter is still usable. If a good amount of light passes through the filter, it is probably okay to reuse (FIGURE D).

5. CLEAN OUT THE AIR BOX.
   Remove and wipe out all the dirt on the underside of the air box. Check the opening for any big obstructions. Small animals sometimes try to make a home in the air box, which is another reason to check your filter regularly (FIGURE E).

6. INSTALL THE NEW FILTER.
   Installing a new filter is the reverse of taking it out. Check to make sure the pleats are pointing in the correct direction and that the filter sits properly in the filter box. Most filters slide in by hand.

TIP

You should replace your air filter as recommended by the manufacturer, even if it looks clean. The paper in the filter can begin to break down from exposure to dirt and moisture, and regular replacement will prevent unwanted debris from entering your engine.

How to: Inspect and Replace the PCV Valve

The positive crankcase ventilation (PCV) valve is a one-way valve that allows fumes and pressure from the engine to be drawn into the combustion chamber of the car and burned, reducing emissions.

WHAT YOU NEED

- Gloves
- Replacement PCV valve (if needed)

If the PCV valve sticks closed, pressure and fuel will build up in the crankcase, and the gaskets and seals of the engine can fail. If it sticks wide open, the car will be pulling too much from the crankcase and the car probably won't accelerate or run well at high speeds.

A

1. LOCATE THE PCV VALVE. The PCV valve is usually on top of the engine, in the valve cover area. It looks like a round plastic tube attached to a flexible hose leading to the intake (FIGURE A).

2. INSPECT THE PCV VALVE. Pull the valve out of its rubber grommet or carefully twist the PCV out of its opening and inspect the underside. Excessive oil buildup or a milky oily substance may indicate a problem with the engine or there may be moisture or coolant getting in the oil.

TIP

If you find excessive oil, dirty oil, or oil that has been mixed with moisture or coolant (it will have a milky look to it), but the PCV seems to be functioning, you may have an internal problem with your engine. Have a professional check it out.

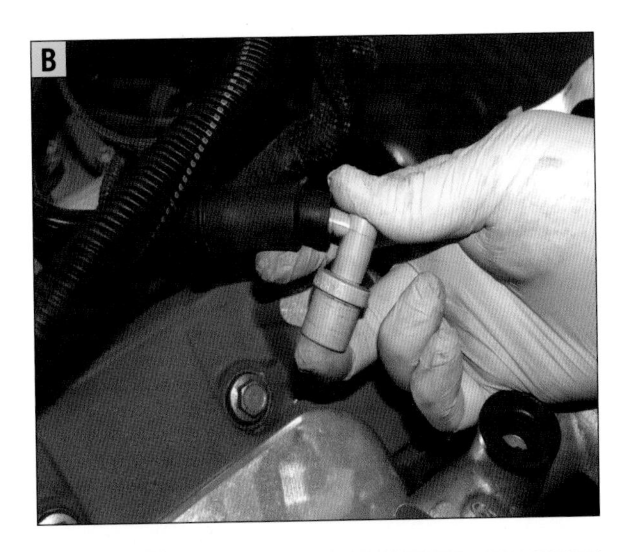

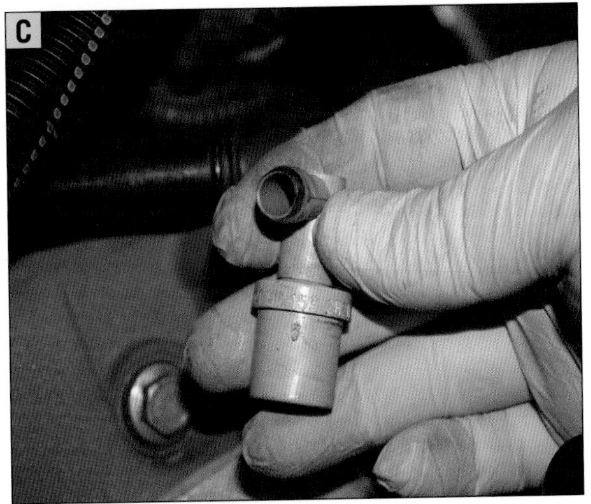

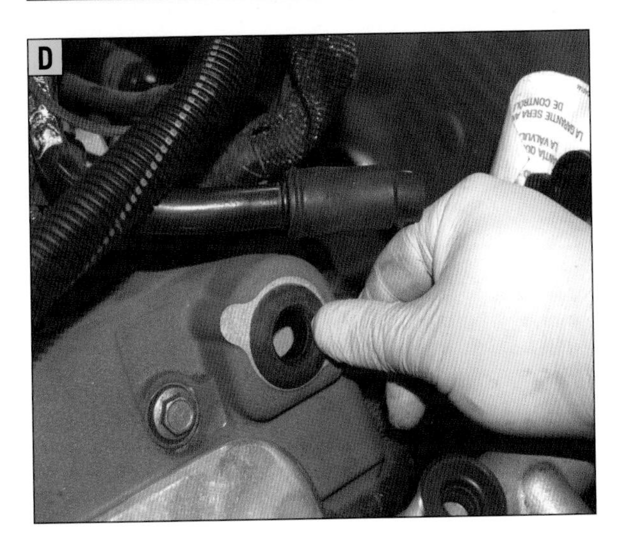

3. VACUUM TEST THE VALVE.
   Turn on the engine and let it idle. With gloves on, place your finger over the opening in the bottom of the PCV valve (FIGURE B). If you feel a vacuum on your finger then the valve is working properly at idle. If you don't get a vacuum, either the valve is stuck closed or there is a blockage somewhere in the lines going to the intake. Check the lines for a blockage if this occurs.

4. SHAKE TEST THE VALVE.
   Carefully pull the PCV valve from the rubber connection hose. It may have a locking connector that needs to be removed depending on the car manufacturer. Give the PCV valve a gentle shake—you should be able to hear the valve rattle inside (FIGURE C). If you hear a rattle then the valve is free and should be working properly. Change the PCV valve if it doesn't rattle.

5. INSPECT GROMMET AND HOSES.
   Check the grommet that the PCV valve sits in for damage. If it has an O-ring, check the ring for wear or damage (FIGURE D). Check the hoses for cracks and leaks, and replace any hoses that are showing wear or not sealing properly.

6. REPLACE THE PCV VALVE
   If you need to replace the PCV valve, pull the valve from its hose, insert the new valve, and reinstall the valve in the grommet or port. If your car has a heated PCV valve, you will need to disconnect the electrical connector from the valve before removal.

THE IGNITION SYSTEM

IN THIS CHAPTER

How the Ignition System Works

Common Ignition System Problems

Spark Plug Location

How to: Change the Spark Plugs

How to: Change Spark Plug Wires

How to: Test for Spark

How the Ignition System Works

Any kind of fire requires three things: fuel, air, and an ignition source. With the exception of electric vehicles and diesel engines, most engines use a spark to ignite the fuel and air inside the engine.

The ignition system in your car is designed to ignite the air-and-fuel mixture at the right time, when the engine has compressed the mixture and it is ready to explode and drive the engine in a rotating motion. To do this, it uses a high-voltage spark that is timed to the engine through the computer or, in older cars, through a mechanical device called a *distributor*.

Spark Plug Spark plugs generate the spark that ignites the air and fuel in the combustion chamber. This voltage is measured in tens of thousands of volts, so you want to be careful around spark plugs while the engine is running. Some engines have more than one spark plug per cylinder to help burn the fuel and air more efficiently.

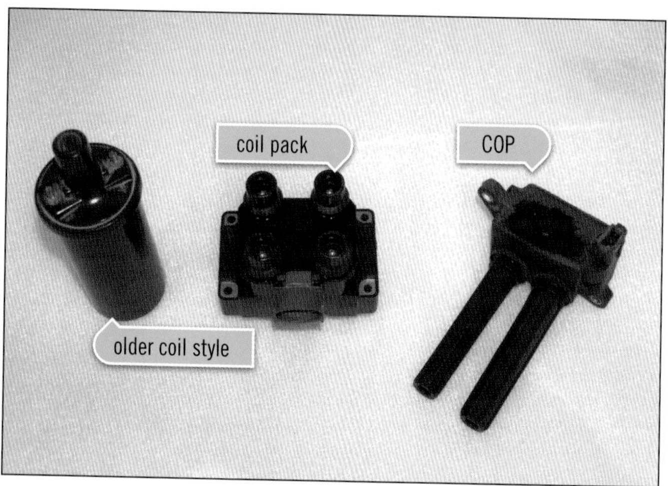

Ignition Coil The ignition coil builds up an electrical charge, as much as 50,000 volts, to trigger the spark in the spark plug. Older, non-computerized cars use a coil that requires a distributor to help send the voltage to the right spark plug. Newer cars use a coil pack—the computer tells it which cylinder to send the spark to. The coil-on-plug (COP), sits right on the individual spark plug and only fires when that spark plug(s) needs power.

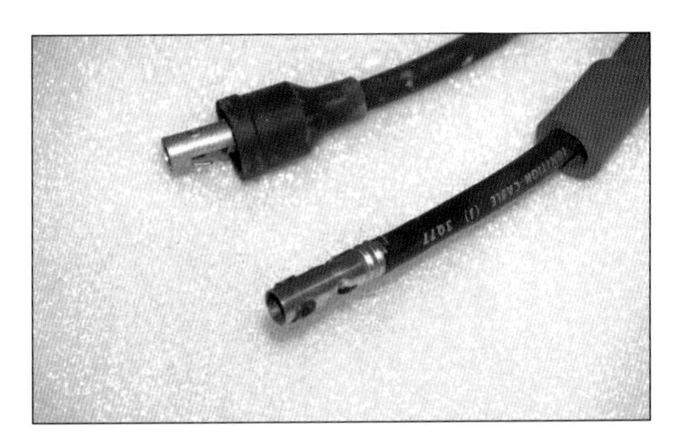

Spark Plug Wires The spark plug wires are responsible for transferring the high voltage from either the coil pack or the distributor to the spark plug. Typical wire construction uses a metal contact that snaps over the end of the spark plug to make a good connection.

Distributor The distributor is responsible for sending (or distributing) the high voltage from the ignition coil to the spark plugs in the correct order. Older vehicles use a traditional distributor that is connected to the camshaft through a gear and rotates around, dispensing the voltage to the spark plugs in sequence through the distributor cap.

CPS Instead of a distributor, newer cars use a camshaft position sensor (CPS). It does the same thing as the distributor, but it sends a signal to the computer when it senses the engine is in the right place, and the computer then knows when to trigger the coil or coils. The computer may also use a crankshaft position sensor along with the camshaft sensor to further refine when to trigger the spark plug.

DIESEL ENGINES

Diesel engines do not use spark plugs. Instead, once the engine is warmed up, they ignite due to the the heat of the compressed fuel, which is under much greater pressure than in a normal gas engine. When the engine is cold, diesel engines use glow plugs to help trigger ignition. A timer or computer turns off the glow plug when the engine can run on its own.

Common Ignition System Problems

Most ignition problems fall into two categories: either the ignition doesn't trigger at the right time (a timing issue), or something is preventing the spark from having a full charge (a resistance or failure of the system).

In the old days, it wasn't possible to detect ignition problems until the car began running poorly. In a modern engine, the computer can sense ignition problems. It uses oxygen sensors to evaluate the air-to-fuel ratio of the engine output. Some engines also have a sensor to detect if the engine is firing too soon (anti-knock sensor), and to tell if the engine is misfiring by checking the motion of the crankshaft.

TIMING ISSUES

Timing issues occur when the spark plug is firing too soon or too late. Modern engines use the computer to trigger the timing based on the sensors checking the crankshaft and camshaft. If the timing in a new engine gets out of sync, it is usually due to the belt or chain wearing out or slipping. A dirty or failing crankshaft or camshaft sensor can also cause timing issues.

On older engines, the distributor and ignition need to be synced, or timed, to the position of the crankshaft and camshaft. This can be done with a strobe light, and by rotating the distributor.

Engine Knocking Knocking occurs due to *pre-ignition*—when the fuel and air in the cylinders ignites before the engine has compressed it to its maximum. It sounds like someone tapping the inside the engine with a hammer and it usually happens under acceleration or when the spark is advancing to give more power.

Pre-ignition is often caused by using gasoline with an improper octane rating, but it can also be due to deposits in the engine heating up and causing the fuel to ignite before the plug fires.

Knocking can be extremely damaging to an engine. Switching to a better fuel or a fuel that can clean out deposits might help eliminate pre-ignition. If not, check your computer codes and plugs for signs of problems.

Spark Not Advancing When a car accelerates, the ignition "advances" the spark, which means it makes the spark turn on sooner to allow the fuel more time to burn as the engine spins faster. Before electronic ignition, vacuum and mechanical weights were used to change the position of the spark. The computer now adjusts the advance as it senses the engine speeding up. Problems with advance in a new engine are the same as timing problems: the computer can't sync properly with the crankshaft and camshaft, or one of the other sensors is giving it wrong information.

Computer Codes The computer can't tell you why the engine is misfiring, but it can tell you that something is wrong with one or more cylinders. When a cylinder misfires, the computer will sense a change in the speed of the crankshaft. It knows which cylinder was supposed to spark, so it can send an error code to tell you which one is misfiring (or if there is a random misfire through all the cylinders). When the computer reports a misfire, the best thing to do is to remove the spark plug (or plugs) and inspect them for problems.

SPARK PROBLEMS OR FAILURE

If the spark plugs don't get a full charge of electricity, they may not trigger a spark big enough to get a good burn of the fuel, or they may not fire at all. The spark plug needs to make a good, clean electrical connection from the coil, through the wire (if used), and ground to the cylinder head. There are lots of things that can prevent proper spark from occurring.

Arcing Worn spark plug wires or signal wires can cause "arcing," which is when the electricity used to fire the spark plugs is dissipated through an arc point. The spark plugs are carrying tens of thousands of volts and it is not hard for a small rub on a wire to cause a voltage arc. An arc, or short, can also ignite fumes outside the engine, so they can be dangerous. If you suspect a faulty wire, look for points where it may be touching the engine, or look for cuts or blackened outer areas where it might be arcing.

Ignition Coil Not Firing The ignition coils generate the voltage that triggers the spark plugs. The number of coils your car has may determine the severity of your problem. If a single coil fails, the engine stops running. A coil pack may have one or all of its coils fail, and coil-on-plug only affects one spark plug.

Grounding Problems If the spark plug is not making good connection to the cylinder head, it can't complete the circuit, and won't fire. If any part of the system loses its connection to ground, the plugs won't fire. Ground issues can be maddening to locate and fix.

Contamination There are a lot of things that can contaminate the spark plug and prevent it from firing: oil from an engine that is seeping into the cylinder, coolant entering the cylinder, deposits from poor gas or poor combustion. Any foreign material that gets between the two points on the plug can keep it from firing normally. Keeping your engine maintained properly will prevent most problems.

Worn Plugs Spark plugs will wear down under normal use. The two points that create the spark will slowly move away from each other, and the "gap" will increase, making it harder to create the proper spark. A plug that is damaged will not create a proper spark.

Spark plugs are designed to operate in a specific heat range, and engines are designed to use a specific plug. In general, you shouldn't need to change heat ranges on your plugs.

SPARK PLUGS

Pulling and inspecting the spark plugs is the best way to find out what is happening inside your engine. They can tell you if there is a problem in one or all of the cylinders of your engine.

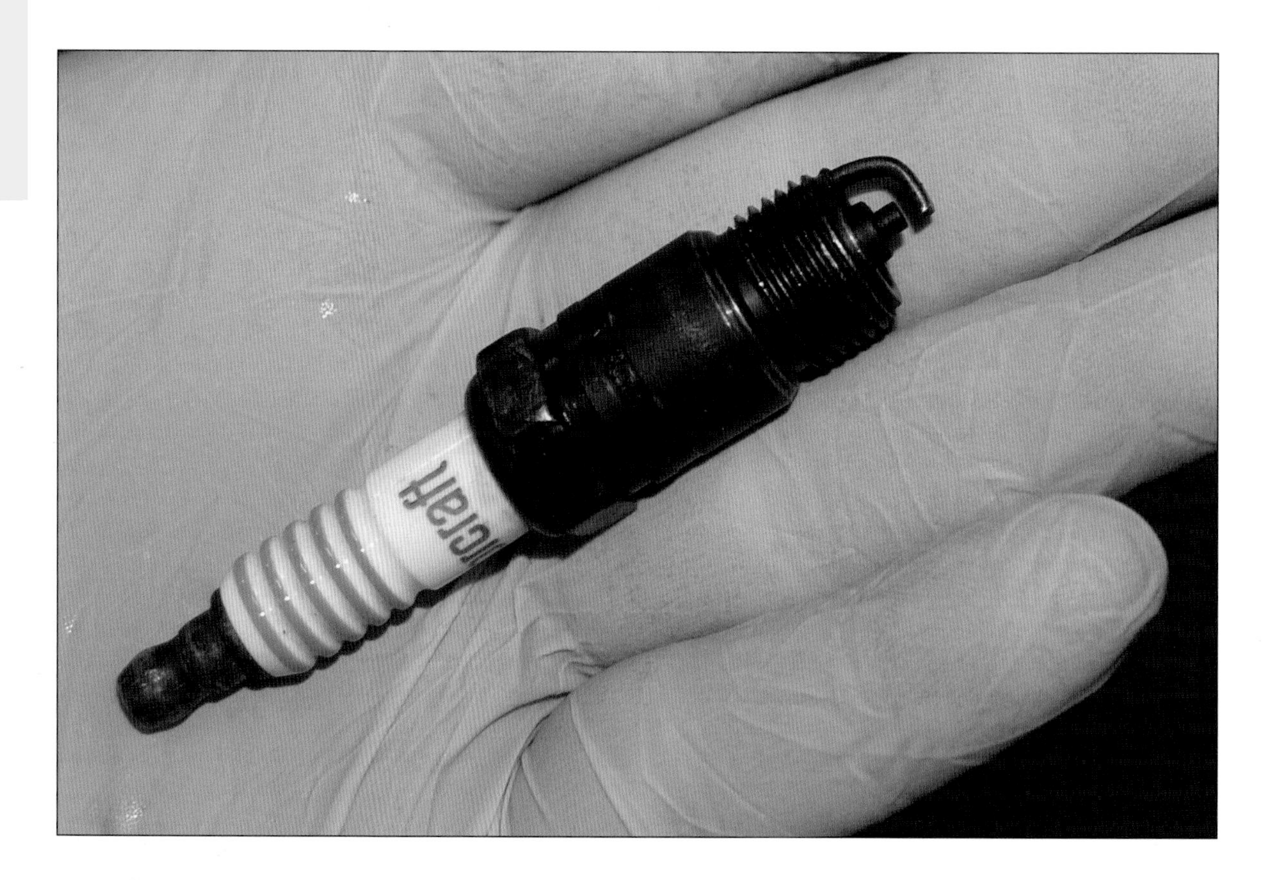

The spark plugs in your engine should look like this one: a light brown or light black color. The points, or electrodes, should be complete and not worn, and the base should not be coated or cratered. All the plugs should come out looking the same.

If you find a plug coated with residue or with burnt, broken, or worn parts, you need to check it against a spark plug chart.

There are dozens of possible spark plug problems. A spark plug chart can help you identify your specific issue. Shop manuals have small charts that show some of the more common spark plug failures, but online charts provided by the spark plug manufacturer or engine builder will usually be more comprehensive.

If your engine starts to run poorly, or if an error code tells you the engine is misfiring or has poor combustion, pull the spark plugs and check them. It is the best way to see what is going on inside the engine.

mon spark plug conditions

NORMAL

Symptoms: Brown to grayish-tan color and sligh electrode wear. Correct heat range for engine an operating conditions.

Recommendation: When new spark plugs a installed, replace with plugs of the same heat ran

WORN

Symptoms: Rounded electrodes with a small amount of deposits on the firing end. Normal color. Causes hard starting in damp or cold weather and poor fuel economy.

Recommendation: Plugs have been left in the engine too long. Replace with new plugs of the same heat range. Follow the recommended maintenance schedule.

CARBON DEPOSITS

Symptoms: Dry sooty deposits indicate a rich mixture or weak ignition. Causes misfiring, hard starting and hesitation.

Recommendation: Make sure the plug has the correct heat range. Check for a clogged air filter or problem in the fuel system or engine management system. Also check for ignition system problems.

ASH DEPOSITS

Symptoms: Light brown deposits encrusted on the side or center electrodes or both. Derived from oil and/or fuel additives. Excessive amounts may mask the spark, causing misfiring and hesitation during acceleration.

Recommendation: If excessive deposits accumulate over a short time or low mileage, install new valve guide seals to prevent seepage of oil into the combustion chambers. Also try changing gasoline brands.

OIL DEPOSITS

Symptoms: Oily coating caused by poor oil control. Oil is leaking past worn valve guides or piston rings into the combustion chamber. Causes hard starting, misfiring and hesitation.

Recommendation: Correct the mechanical condition with necessary repairs and install new plugs.

GAP BRIDGING

Symptoms: Combustion deposits lodge between the electrodes. Heavy deposits accumulate and bridge the electrode gap. The plug ceases to fire, resulting in a dead cylinder.

Recommendation: Locate the faulty plug and remove the deposits from between the electrodes.

MECHANICAL DAMA

Symptoms: May be caused by a foreign object in the combustion chamber or the piston striking an incorrect reach (too long) plug. Causes a dead cylinder and could result in piston damage.

Recommendation: Repair the mechanical damage. Remove the foreign object from the engine and/or install the correct reach plug.

Spark Plug Location

Spark plugs are always located on the cylinder head, but the position can vary. They can be on the top or on the side of the engine, they may be covered by a spark plug wire or a coil, they may be under a cover, and there could be more than one spark plug per cylinder.

Check your owner's manual to see how many cylinders your engine has, or copy down your Vehicle Identification Number (VIN) and take it to an auto parts supplier. They can tell you how many cylinders your car has and where the plugs are located.

Let's take a look at where the spark plugs are located on different types of engines.

This muscle car from the 1960s uses a mechanical distributor and spark plug wires. The big engine in the small car makes the middle spark plugs difficult to change.

This Toyota Corolla has four spark plugs and they are located right on top of the engine. Their power comes from an electronic distributor and spark plug wires. With an extension and spark plug socket, anyone can change these plugs.

The spark plugs are not visible because they're hidden beneath the fuel rails and intake system.

This mid-90s Mercury uses two coil packs and plug wires to fire the plugs. The passenger side plugs are fairly accessible; the driver side plugs are buried under the engine intake system and some other hoses. This requires a little more effort to change.

Here is a coil-on-plug (COP) engine. Don't be intimidated by the coils—they come right off and the plugs are easy to get to. These plugs are easy to change due to their location.

These spark plugs are hidden under a cover.

This V6 front-engine drive has three plugs on the back side of the engine. Changing these requires unhooking the motor mounts and pulling the motor forward. Not a job for a novice!

How to: Change the Spark Plugs

Changing your own spark plugs can be easy or difficult, depending on where the plugs are located. Make sure you have access to your spark plugs before you begin.

WHAT YOU NEED

- Spark plugs
- Socket, wrench, and extension (if needed)
- Spark plug gap tool
- Anti-seize compound
- Torque wrench
- Tools to remove the coil-on-plug (if needed)
- Safety gloves

1. LOCATE YOUR SPARK PLUGS.
   With the engine cold, locate the plugs and determine whether you can reach them easily. Even if you can see them, there may be other things in your way that will prevent you from getting to all the plugs (FIGURE A).

SPARK PLUG SOCKETS

Select a socket that is meant for changing plugs and is the appropriate size for your vehicle. Sockets come in the following sizes: ⅝ inch, 13⁄16 inch, 16 millimeters, and 18 millimeters. Some have a rubber boot inside to keep the spark plug in place when installing or removing, and some are magnetized. You'll also need a matching socket wrench. Depending on the location of the plugs, you may need a socket extension (a bar that allows the sockets to extend away from the wrench and reach down in holes) and perhaps a swivel (part that enables the socket to rotate at an angle to get around obstacles).

2. REMOVE THE SPARK PLUG.
Pull one spark plug wire at a time. The wire should pull off by hand, or you can use special spark plug wire pliers. If you have coil-on plugs, unscrew one coil at a time and disconnect the electronic connector. With the wire or coil out of the way, use the spark plug socket and wrench to remove the plug (FIGURE B). Once the plug is out, inspect it for any kind of damage or abnormal wear. Don't pull them all at once—if you find one with a problem, you may not put it back in the correct cylinder.

3. CHECK THE GAP ON YOUR NEW PLUGS.
Before installing your new plugs, check that the gap between the two electrical points is correct. Using a spark plug gap tool or feeler gauge, insert the tool between the post and metal L-shaped bar. The tool should slide in and out but not have any space between the tool and metal points (FIGURE C). If you need to adjust the distance, you can very gently bend the L-shaped bar in or out and retry the gap.

4. APPLY ANTI-SEIZE GREASE.
Dab a little anti-seize compound on the plugs before installing them in the engine. This will help to prevent the plugs from becoming stuck and stripping the threads when they need to be removed (FIGURE D).

5. INSTALL THE NEW PLUGS.
   Insert the new plug into the engine using the spark plug socket and tighten by hand (if possible) until the washer contacts the engine. Using your torque wrench, tighten the plug to what your car builder recommends. The torque spec can be found in your shop manual or online (FIGURE E).

6. RECONNECT THE SPARK PLUG WIRE.
   Listen for a "click" when installing the wire to let you know the wire is seated properly (FIGURE F). If you have coil-on-plug, reattach the connectors keeping the coil in place, and re-connect the computer connection if you unhooked it. You can now proceed to the next plug.

After changing all the plugs, start the engine and listen for misfires. The most common problem when changing the spark plugs is plug wires not seating properly, or if you pull them all at once, getting a wire crossed.

TIPS FOR CHANGING SPARK PLUGS

The spark plug covers, or "boots," can get stuck on the spark plugs, and sometimes they are hard to get to. These special spark plug wire pliers grab the boot gently and help remove stuck-on plug wires.

Likewise, sometimes the plug wires don't want to snap on the plugs, and you can't move the boot. A little spray lubricant usually helps pull the boot down the wire, and when you are sure the end is attached, it slides right back down the wire and in place.

If your spark plug socket does not have a rubber insert or is not magnetized, it can be difficult to install the plug in the hole. Use a piece of rubber hose that fits tightly around the plug to snake it down into position and get the threads started.

How to: Change Spark Plug Wires

Changing plug wires is similar to changing the spark plugs—the less accessible the wires are, the more difficult the job will be. Sometimes the wires are in a place you just can't access. Cars with coil-on-plug (COP) technology will not have wires, but an individual coil for each spark plug. If the coil fails, it can be changed.

WHAT YOU NEED

- New spark plug wires specific to your car.
- Spark plug boot tool (optional)
- Spark plug grease (optional)
- Gloves and eye protection

A

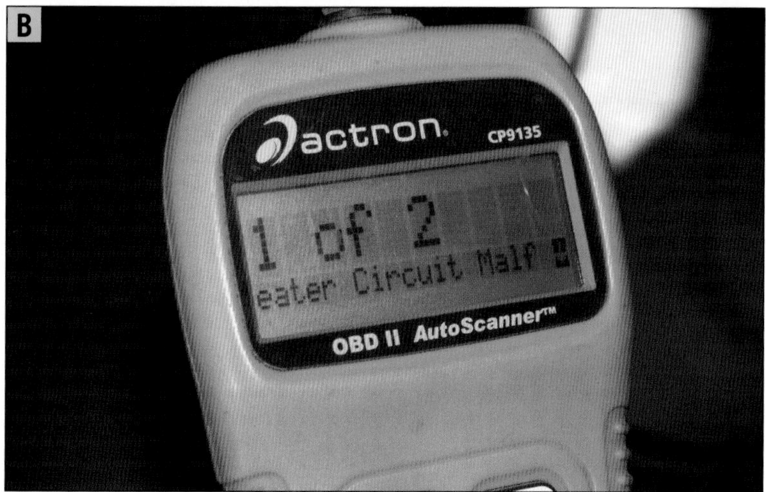

B

1. PREPARE TO CHANGE THE WIRES.
   Wait until the engine is cold. Before you start tearing off plug wires, take a photo of how things are routed on the engine for future reference (FIGURE A).

2. CHECK FOR ERROR CODES.
   Use a code reader to check for engine misfire codes. Note if any cylinders are misfiring and then clear the misfire codes from the reader (FIGURE B).

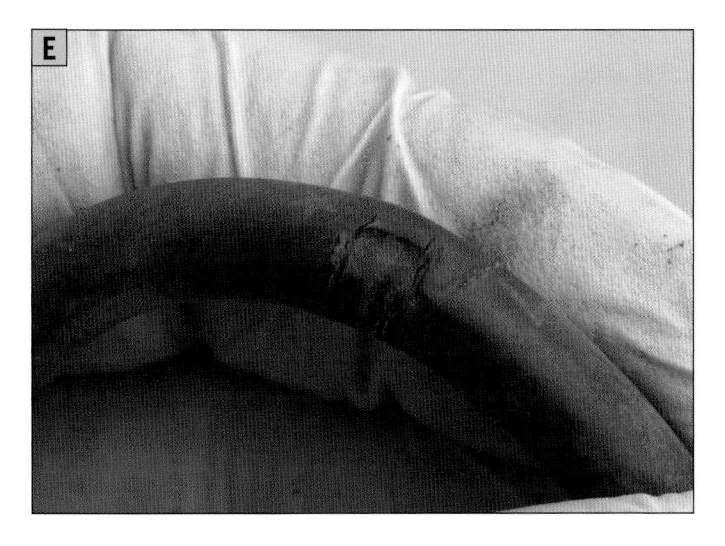

3. PULL THE PLUG WIRE.

You can start at the spark plug or coil end, but pull the plug wire and start removing it from any straps that are holding the wires together. On this vehicle, the location of the wires under the fuel rail makes it necessary to use a spark plug boot tool (FIGURE C).

On this coil pack, the connectors are squeezed together and the end comes off. Your wires may connect differently (FIGURE D).

Always pull one wire at a time to avoid installing the wire on the wrong plug. This also allows you to test after each wire if you wish to see if the wire was installed properly.

4. INSPECT THE WIRE.

Look for corrosion on the contact ends, brittle points, cuts, rubs, or abrasions that could cause sparking. This wire experienced insulation failure and was arcing against the valve cover (FIGURE E). If you find a worn spot, check the point where it was rubbing and look for evidence of arcing, such as black marks on the metal. Try to route the wire differently to avoid contact with metal or heat sources.

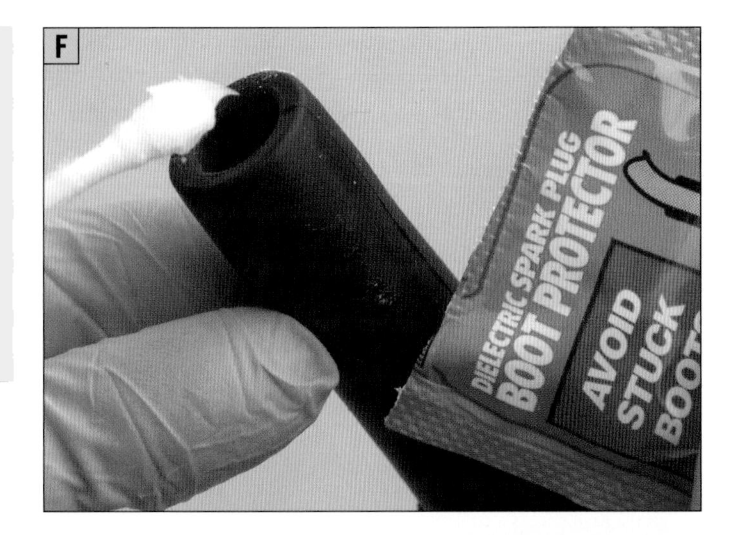

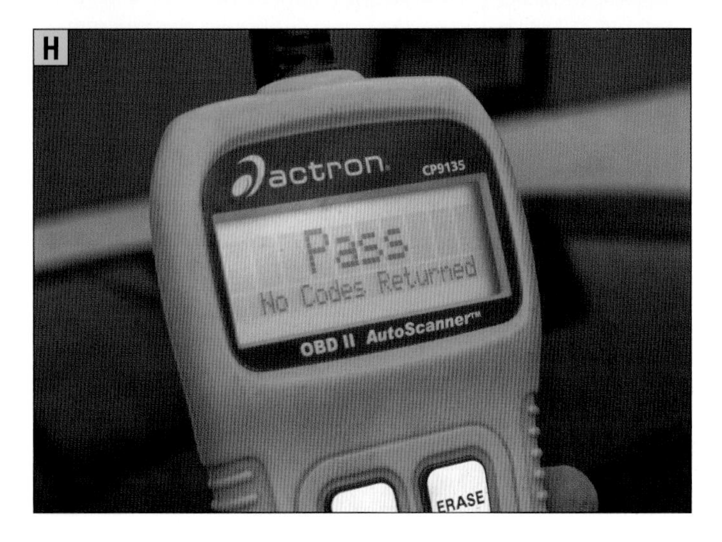

5. APPLY GREASE TO THE PLUG ENDS.
   Applying dielectric grease to the inside of the spark plug boot prevents the boot from sticking to the plug and keeps water out of the connection (FIGURE F).

6. INSTALL THE NEW WIRE.
   Starting at either end, install the new plug wire. When installing the coil end, make sure it snaps in place and if your wire has a locking device, make sure it locks closed. On the spark plug end, listen for it or feel it click in place over the spark plug (FIGURE G).

   Once you have the wire in place, you can move on to the next wire until all wires are exchanged.

7. RECHECK YOUR COMPUTER CODES.
   Once you have all the wires changed, start the engine and let it run for a while. Recheck your engine computer codes (FIGURE H). If you have a plug wire not making good contact, your engine may post an error code. If you are getting error codes and you know your plugs and wires are seated properly, you may have another problem in your ignition system.

COIL-ON-PLUG SYSTEMS

Changing the coils in a coil-on-plug system is similar to changing regular spark plug wires. On this engine, we removed the cover on the top of the valve cover to expose the coil-on plugs. Sometimes the coils are screwed in place, but on this car we simply pulled the coil off of the plug. Inspect the insulator, looking for damage and corrosion. Since the connection to the spark plug (a small wire or an insulator) is usually connected permanently to the coil, if you find damage you will need to change out the whole coil.

How to: Test for Spark

If you suspect that one of your cylinders is misfiring, you need to find out if a bad spark could be the cause. The first step is to check your computer codes. If the computer detects a bad combustion burn, it will post an OBDII code to tell you which cylinder is causing a problem (or if they all are). Misfires can be caused by an improper mixture of fuel and air, loss of compression, bad timing, or bad spark.

WHAT YOU NEED

- Code reader
- Inductive timing light or tester
- Shop manual

When you test for spark, you're not checking or setting the timing; you're just making sure that the spark plug and spark plug wires are good. You can use an inductive timing light to determine if electricity is flowing through the plug wires properly. .

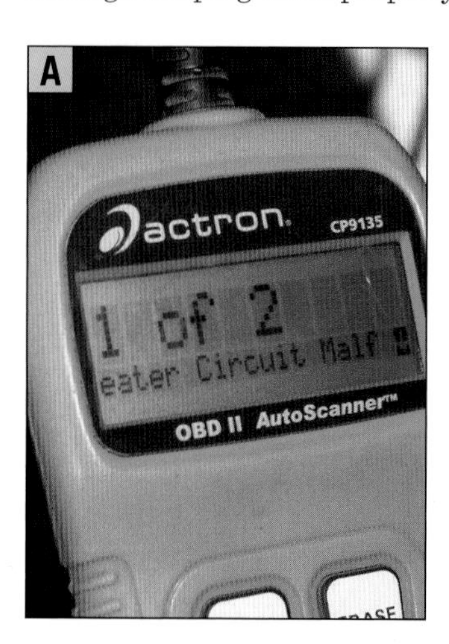

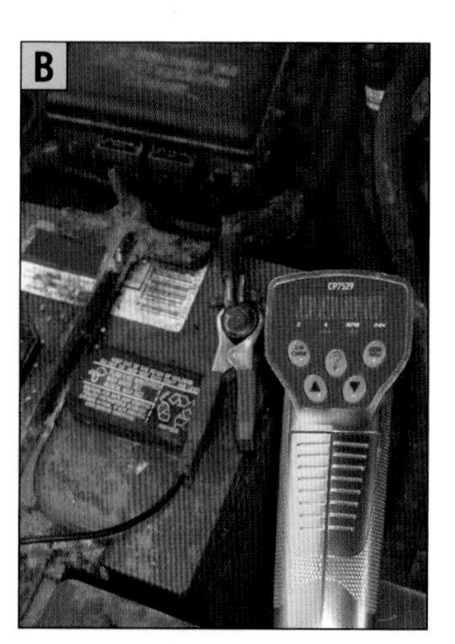

1. CHECK FOR ERROR CODES.
   A code reader will tell you which cylinders are misfiring. Write down which cylinders have problems and then use your shop manual to see where in the engine the cylinder is located. These are the spots you want to check (FIGURE A).

2. CONNECT THE TIMING LIGHT TO THE BATTERY.
   The timing light connects to the positive and ground terminals on the car battery (FIGURE B).

3. CONNECT THE INDUCTIVE CLAMP TO THE SPARK PLUG WIRE.
   Place the inductive clamp on the suspected spark plug wire. It can be anywhere along the wire, just make sure it is clear of moving or hot engine parts (FIGURE C).

4. TURN ON THE ENGINE AND ALLOW IT TO IDLE.
If your engine won't run, try turning the engine over enough to see if the light will trigger.

5. CHECK FOR THE SPARK.
With the engine running, the timing light should begin to flash steadily (FIGURE D). If it is not flashing, the spark plug is not triggering. If it is erratic there may be a problem with the circuit.

6. CHECK THE REMAINING WIRES.
Check all the wires and confirm that all the spark plugs are firing at the same rate. Be sure to turn off the engine before disconnecting the timing light and reconnecting it to a different wire. If you find a bad circuit, you may be able to pull the spark plug and check for problems.

COIL-ON-PLUG SYSTEMS

Some coil-on-plug setups don't have a plug wire and are hard to test without some unique tools. When you check the computer codes, the computer should not only tell you if you have a misfire in a cylinder, but also if the coil itself is failing. If the coil has a problem, it will post a unique OBDII error code. If you get a misfire code but not a coil code, it may be the spark plug or something else causing the problem. If you get both codes, it is time to see a professional.

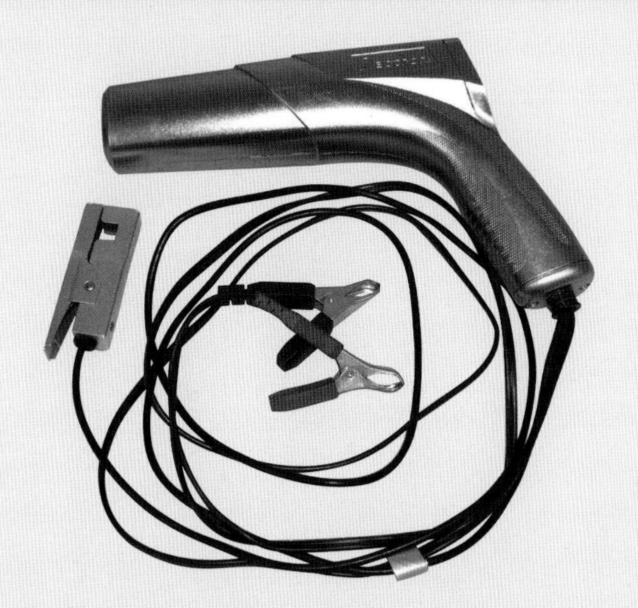

INDUCTIVE TIMING LIGHTS

Timing lights come in many different styles, including this gun-shaped variety, but they all work similarly. The timing light is powered by connecting to the car's battery, so it never runs out of juice. It also has an inductive clamp that connects to the spark plug wire and can detect the jolt of electricity each time the spark plug fires. Mechanics use timing lights to set the engine timing on older vehicles, but they're also useful when checking for a bad spark.

THE COOLING SYSTEM

IN THIS CHAPTER

How the Cooling System Works

Combustion engines generate a lot of heat, and to prevent the metal components from overheating, fuel-burning cars need a way to transfer the heat from the engine to the outside air. This is the job of the cooling system.

Most cars have a liquid cooling system, which uses a liquid coolant that runs through the engine and is then pumped into a radiator to transfer the heat to the air. Car engines run at nearly the same temperature that water boils (212°F or 100°C), so they must use a combination of coolant and water to allow the coolant to run hotter than straight water. In cold temperatures, the same liquid acts as an antifreeze, preventing the water from freezing and expanding, which can crack metal parts.

Here are the common components of a liquid cooling system.

Radiator The radiator is usually mounted on the front of the car. Coolant from the engine is pumped through it and the heat from the coolant is transferred from the tubes to the small fins lining the tubes. As air flows over the fins, the heat is transferred to the surrounding air. The engine then pumps the cooled coolant from the bottom of the radiator back into the engine to pick up more heat.

Water Pump The water pump takes cooled liquid from the bottom of the radiator and pushes it through the engine and then back to the radiator. The water pump may be driven by a belt from the engine, or electrically from the alternator.

Thermostat The thermostat regulates the engine temperature. Since metal expands when it heats, the engine is designed to run best at a heated temperature. If the engine takes too long to come up to temperature it can cause problems. The thermostat senses when the temperature has reached an optimum point and then opens up, allowing the coolant to run through the radiator.

Cooling Fan A cooling fan is used when the car is not moving fast enough to allow air to flow over the radiator. Cooling fans are mostly electric, but they can be powered by an engine belt. Mechanical fans sometimes use a clutch to allow them to spin freely when the car is moving fast enough, saving energy. Electric fans can be turned on and off by temperature sensors or by the computer.

Heater Core The waste heat from the engine is used to heat the inside of the car when needed. It does this using a heater core. The heater core looks like a miniature radiator, and it works just like the radiator by transferring heat to the surrounding air. It is usually mounted in a box under the dash alongside with the air conditioner evaporator.

Hoses Most engines use a combination of hoses and tubes to transfer the heat from the engine to the radiator and the heater core. The flexible hoses allow the engine to move during operation and are usually held in place with clamps.

Overflow Tank An overflow tank, also called a de-gas tank or coolant reservoir, is used as a reservoir for extra coolant. As the engine heats up and cools down, this tank holds the extra coolant, ready to be delivered by the water pump. It also provides a safe way to check the coolant level and add more coolant, if needed.

Coolant The coolant, or antifreeze, is the liquid component of the liquid cooling system. It takes heat out of the engine and prevents the cooling system from freezing in low temperatures. Antifreeze is designed to be mixed with water, so don't run straight antifreeze in your car. It is also highly toxic, but has a "sweet" smell that makes it attractive to dogs and other animals. Keep it away from pets and children.

Common Cooling System Problems

The cooling system tells you it has a problem one of two ways: you find a leak on your driveway, or the engine starts overheating. The cooling system is not maintenance-free and requires regular checks to stay in good condition.

LEAKS

There are a lot of flexible connections in the cooling system and this is where leaks most commonly start. The rubber hoses will begin to wear out and the joints will loosen from movement. A leak can happen anywhere in the system: the water pump can leak, the radiator can rust through, or the cylinder block can lose a gasket. Leaks in hoses can usually be repaired easily, but leaks in big components like radiators and water pumps generally require the attention of a professional.

Leaks that aren't addressed can lead to overheating, which leads to engine failure. Get a leak repaired right away.

RADIATOR FAILURE

The coolant in the radiator is mixed with water, and the water can bring with it minerals and deposits that will clog and corrode your radiator. When this happens, the coolant can't pass through the tubes of the radiator and the engine overheats. If it rusts out, it will eventually drain down and cause overheating.

A mildly corroded radiator can be flushed with commercially available products, or some pros offer a power flush that really gets the radiator clean. You can flush your own radiator, but a severely corroded radiator should be replaced.

WATER PUMP FAILURE

The water pump is subject to failure in several ways. A mechanical pump that is mounted to the engine may fail at the gasket points where it connects to the engine, or it may stop working if the engine belt driving it slips or breaks. An electric pump may fail if its internal motor dies. The bearing in the water pump can also give out. A whining sound may indicate that a water pump is about to fail.

THERMOSTAT FAILURE

The thermostat is responsible for regulating the flow of coolant by measuring the engine temperature and opening if the engine needs to be cooled down. If the thermostat fails while closed or partially closed, the car will overheat because the coolant can't flow. If the thermostat fails while open, it will take the engine a much longer time to come up to temperature, which can be hard on the oiling system and cause oil crystallization on the engine components. Don't run your car without a properly functioning thermostat.

WORN-OUT COOLANT

Coolant needs to be topped off or replaced periodically. You can check the level and condition of the coolant in the overflow tank. If the level is low, it can be "topped off" with distilled water. If it's getting brown with contaminants or rust, it's time to change the coolant.

HOSE CLAMP FAILURE

Hose clamps usually don't fail unless they have been removed and reinstalled. Be careful when reinstalling hose clamps that have been stretched—they can fail on reinstallation.

Locating Leaks

Leaks in the cooling system can be internal or external. Internal leaks, where the coolant is getting into the cylinders and oil pan, can cause serious problems and should be addressed by a professional. External leaks might be easily repaired depending on where they are located.

Here are some ways you can locate a leak in your cooling system.

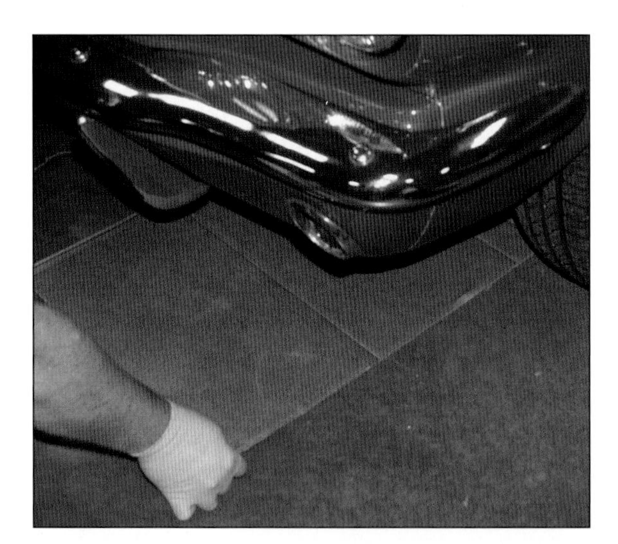

Use a piece of cardboard. If you suspect you have a leak but aren't able to locate the source, place a piece of clean cardboard under the engine area of your car when you park it for the night. In the morning, the position of the liquid on the cardboard should help you find the leak.

Inspect the hoses. Rubber components, like the hoses, can often develop leaks since they are designed to flex. Wearing surgical gloves, gently run your hands around the cold engine and feel for leaks on the back sides of the hoses. Check the hose connections and make sure the hoses are tightly held in place. Also check for rusty water or stains on the hoses. As you go, check your gloves for the fluorescent color of coolant.

Inspect the radiator. The radiator is made of thin metal fins, which can easily be punctured or rusted. This radiator suffered from a cooling fan that dug into the middle of the radiator fins. Corrosion or mineral deposits due to damage could be the sign of a leak.

Inspect the radiator cap. The rubber seal in the radiator cap can fail, and the metal backing holding the seal can crack and allow pressure to be released. When the engine is cool, inspect the area around the radiator cap for signs of escaping coolant. Remove the cap and inspect the underside. Look for a milky, oily substance—this may be a sign of an internal coolant leak.

Check the heater core. The heater core is located inside the passenger compartment, and if it starts to leak, it may soak the carpeting with coolant, or it may weep out of the air conditioner drain. If the inside carpet is getting wet, blot it with a clean paper towel and look for coolant color or rust.

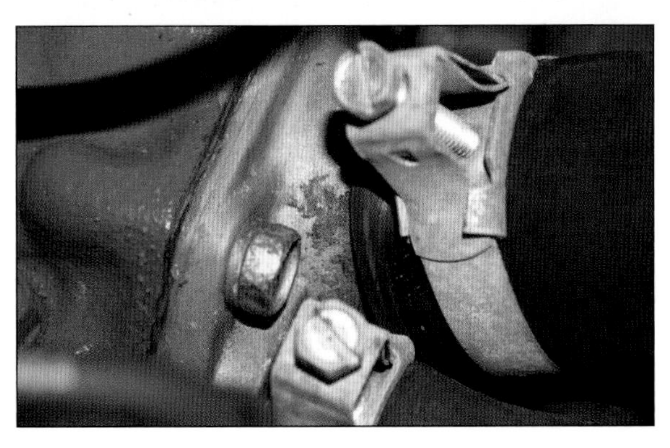

Check the thermostat housing. Some thermostats are mounted on the top of the engine where the hot coolant is returned to the radiator, and this is a key area for leaks. You can see evidence of a past leak here from the mineral and rust buildup on the aluminum housing, and the re-positioned hose clamp.

Inspect the engine block freeze plugs. The engine is equipped with something called *freeze plugs,* which pop out if the water in the engine begins to freeze and expand. These freeze plugs can rust out or weep. Look for signs of rust trailing down from the freeze plugs.

Check the head gaskets. There is a gasket between the cylinder block and the cylinder head made of metal. If it fails, coolant may flood into the cylinders or it may leak out the side of the gasket. When this happens, you may see a small stream of fluid running down the length of the engine.

Check the water pump. A water pump failure can happen at the gasket between the pump and engine or at the front bearing on the rotating shaft. The water pump may have something called a *weep hole,* which is a hole near the rotating shaft. This weep hole is used if the water pump is beginning to fail. If the weep hole is leaking oil, it means the front bearing is failing. If it is leaking coolant, it means the seals in the pump are failing. If you see anything coming out of the weep hole, have it checked right away.

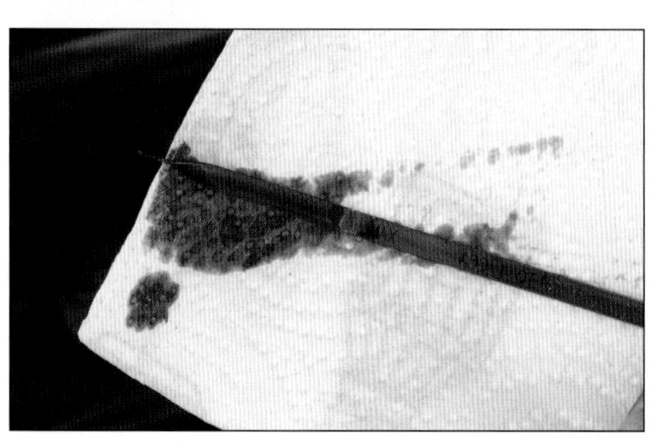

Check the oil. Remove the engine oil dipstick and look for a white, milky substance. This is a sign that there is an internal engine leak and you need to have it checked.

Pressure test the system. Use a radiator pressure tester to test the system. This is a small pump and gauge that attaches to the radiator in place of the cap. Bring the cooling system up to pressure and monitor the gauge for a couple of minutes. If the system doesn't lose too much pressure, there aren't any serious leaks. A quick drop of pressure may confirm an internal leak that you can't see. Pressure testers can be expensive, but you may be able to rent one from an auto parts store.

FIXING LEAKS

With the exception of tightening or replacing flexible hoses, most cooling system leaks are difficult to repair and should be left to a professional. Temporary fixes are unreliable, and you may make a bad problem worse.

Be wary of temporary fixes such as radiator "stop leak" products. These products are made from either a very fine metal powder and binder or a fluid that swells to fill the seals and gaskets. The solution is pumped through the cooling system and it plugs up the small holes.

These products can be used for a short-term fix, but they aren't meant to be a permanent solution. Don't use these products and consider your problem solved. If you have corrosion, it will continue to get worse. The extra powder floating around in your coolant can clog something else, and if you flush your cooling system, you will probably reopen the leak. The best solution is to fix the leak properly.

How to: Check and Add Coolant

You should check the level and condition of your coolant on a regular basis, usually at the same time as your oil change. This is especially important if you frequently drive on dirt roads, as it is easy for dust to get into the coolant system and turn into a muddy mess. Rust, mineral deposits, and dirt in the radiator can reduce the ability of the radiator function and may cause major problems.

WHAT YOU NEED

- Paper towel
- Coolant

1. CHECK COOLANT LEVEL AND CONDITION.
   Look for marks on the outside of the reserve tank indicating the proper levels and check that the coolant level is correct. Check the condition of the coolant in the tank by removing the fill cover and inspecting the contents. Look for dirt or mud, rusty coolant, or other contaminants (FIGURE A).

2. TOP OFF THE COOLANT IF NEEDED.
   If your coolant level is low but in good condition, remove the fill cap and pour in coolant to the proper level (FIGURE B).

CAUTION

Coolant systems build pressure when hot, and this pressure can blow scalding hot coolant out and cause severe injury. Never open a radiator cap while the system is hot.

C

D

3. REMOVE THE RADIATOR CAP.

With the engine cold, remove the radiator cap. The cap comes off in two stages. Push down on the cap and rotate it counter-clockwise and it will stop in a middle point. This is a safety point that allows pressure to be released without coolant spraying straight out. Push down again and turn the cap the remainder of the way until you can pull it off.

4. INSPECT THE RADIATOR CAP.

Inspect the underside of the cap for contaminants, rust, or cracking in the rubber seals or metal body (FIGURE C). If you see any wear on the cap, it should be replaced. Use a paper towel to clean off any debris on the seals of the cap.

5. INSPECT THE RADIATOR.

Look down in the radiator and check the condition of the coolant and inside of the radiator (FIGURE D). Look for scale buildup on the coolant tubes inside. If you see rust and buildup, you should flush your coolant system, even if it has not reached the manufacturer's recommended change point.

If you see a white, milky, oily substance, you may have a problem in the engine that is causing oil to get into the coolant. If you find a lot of this material, take it to a professional for inspection.

6. REINSTALL THE CAP.

Be sure to push down and twist the cap past the safety point and until it completely stops.

How to: Flush and Fill the Cooling System

The fluid in the cooling system needs to be changed regularly, but the frequency differs by manufacturer. Check your owner's manual and follow your manufacturer's recommended schedule. Before you fill the cooling system, flushing it by running water through it will help to clear rust, scale, and dirt.

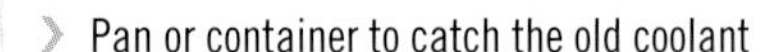

- Pan or container to catch the old coolant
- Funnel
- Paper towels or clean rags
- Safety gloves and glasses
- Garden hose
- Pre-mixed coolant, or coolant and distilled water
- Flushing additive (optional)
- Tools to remove the reserve tank (if needed)
- Jack and jack stands (if needed)
- Extra liquid containers for recycling (if needed)

1. LOCATE THE RADIATOR DRAIN.
   The coolant drain will be at the lowest point of the cooling system, usually in the bottom of the radiator. This one turns counter-clockwise and pushes in, the opposite of a typical water valve (FIGURE A). Some turn part way and then pull out.

FLUSHING ADDITIVES

You can flush your cooling system using straight water, or you can buy flushing additives that will help clean the crusted deposits out of the cooling system. These products are very corrosive, so use extreme caution when handling them.

TYPES OF COOLANT AND MIXES

You can choose to fill your cooling system with a full-strength coolant mixed with distilled water or you can purchase a pre-mixed coolant. In most conditions, the ratio of coolant to distilled water should be 50:50. Check your owner's manual to see if a specific brand of coolant is recommended.

B

C

D

2. RAISE CAR IF NEEDED.
   Since some cars are very low to the ground, it may be necessary to jack up the front end and place the vehicle on jack stands.

3. DRAIN THE RADIATOR.
   With the engine cold, place your pan under the drain valve and open the drain, allowing the coolant to exit (FIGURE B). Remove the radiator cap to allow air to fill the coolant system.

   Some cars have air bleed valves located in the system. These should be opened when draining the coolant. Check your shop manual and see if your vehicle is equipped with a bleed valve (FIGURE C).

4. FLUSH THE SYSTEM WITH WATER.
   The first round of flushing can be done with a garden hose. Place the hose in the radiator cap opening and then turn on the water, letting it flow through the radiator to flush out loose contaminants. The water and dirt will run out the drain valve. Let the radiator flush for a couple of minutes (FIGURE D).

5. CLOSE VALVES.
   Once you are done, close the drain valve on the radiator and any air bleed valves you may have opened.

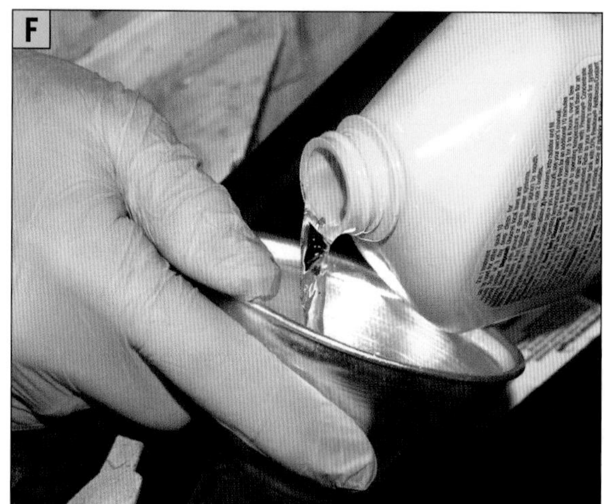

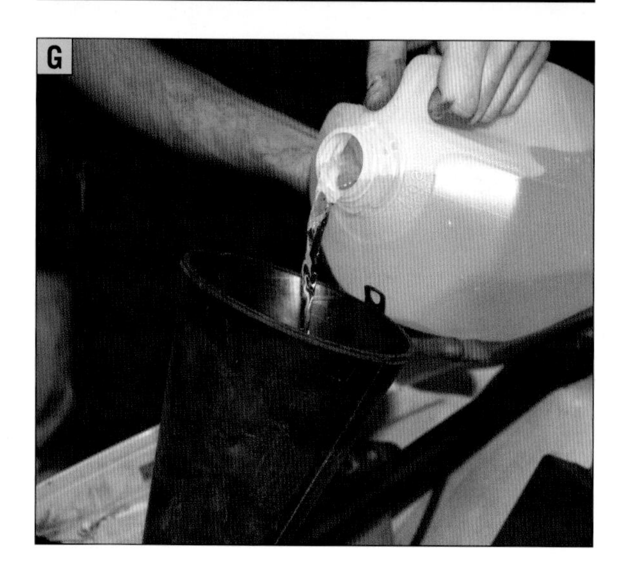

6. **CLEAN THE RESERVE TANK.**
   If you're able to easily remove the reserve tank, clean it out to remove the slime and buildup (FIGURE E). If you cannot easily remove the tank, you can flush the tank by carefully removing the hose that attaches it to the radiator and flushing it with water.

7. **ADD FLUSHING ADDITIVE AND WATER. (OPTIONAL)**
   If you are not using a flushing additive, proceed to step 9. If you are using a flushing additive, add it to the radiator (FIGURE F). Then fill the cooling system with water. When it reaches the top, stop filling and wait a few seconds to allow the system to "burp" air out of the lines. You may need to top off the water several times while the system releases air.

8. **CLOSE THE RADIATOR CAP AND RUN THE VEHICLE.**
   Turn the heater on to make sure the water is flowing through the heater core. Allow the vehicle to come up to temperature, and let it run for about 10 minutes to allow the flush to work through the system. Turn off the car and let the system cool, then drain the flush and water from the cooling system.

9. **FLUSH WITH WATER.**
   If you are not using a flushing additive, or you have completed steps 7 and 8, close the drain valve and fill the system with fresh water (FIGURE G). Bring the car up to temperature as described in step 8. When the car is cooled down, drain the new water out and close the drain. If you have the car on jack stands, you can lower it at this time.

10. ADD THE NEW COOLANT.

Check your owner's manual for the capacity of your coolant system. If you are using pre-mixed coolant, simply pour it in the radiator and burp the coolant until full. Once burped, reinstall the radiator cap and then fill the reserve tank to the cold level.

If you are using a full-strength coolant, mix to your desired ratio with distilled water. It is not necessary to pre-mix the two before filling the system, but put the coolant in first.

RECYCLING OLD COOLANT

Coolant is very toxic and can be fatal if ingested. The flush water is also highly corrosive and contains traces of coolant. These fluids should be recycled at a proper recycling facility. Check your local codes and suppliers for more information on where to recycle these fluids.

How to: Replace a Hose

The flexible coolant lines in your engine allow the engine to move and vibrate while the radiator and heater core remain fixed. These hoses need to be changed if they become brittle or spring a leak. The hose clamps that hold the hoses in place can also loosen and leak, and may need to be tightened or replaced.

WHAT YOU NEED

- Replacement hose
- Pan to catch coolant
- Safety gloves and glasses
- Paper towels or rags
- Tool to remove and install the hose clamps
- Utility knife or razor blade (if needed)

If you're able to see both ends of a leaking hose, you should be able to change it yourself. Be sure to wait until the engine has cooled before draining the cooling system.

1. DRAIN THE COOLING SYSTEM.
   See "How to: Flush and Fill the Cooling System" for information on draining the coolant.

2. LOOSEN THE HOSE CLAMPS.
   Using the correct tool for your clamps, loosen the clamps and pull the hose back away from the connection. There is usually an internal lip on the hose connection, and the clamp must be moved past this lip to release the hose (FIGURE A).

HOSE CLAMPS

There are several different styles of hose clamps. **Compression clamps** use a screw or bolt to draw in the band tightly. **Spring-style clamps** use spring action to hold the hose in place. Most manufacturers use spring-style clamps.

You need a screwdriver or wrench to tighten or remove a compression clamp. Spring-style clamps are opened using pliers. Regular pliers will work, but if you plan to do a lot of work on hoses, you might want to invest in special **clamp pliers**, which have a slot in them to hold the clamp in place.

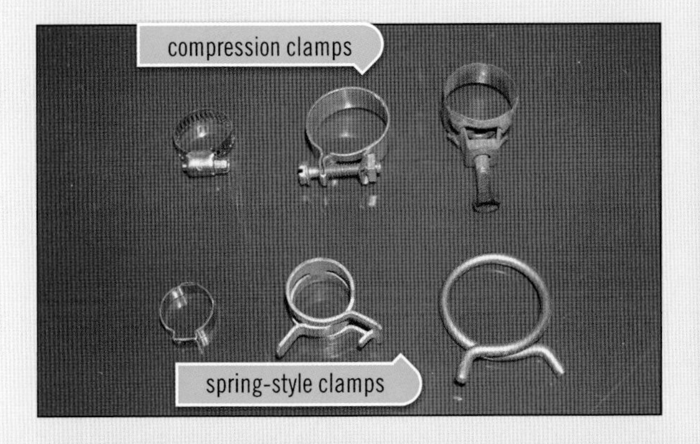

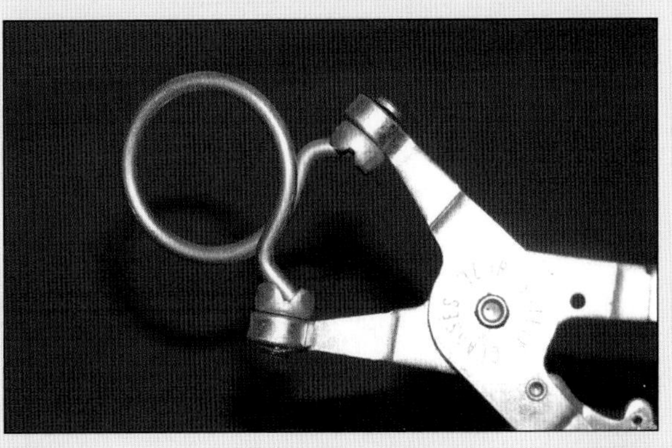

3. REMOVE THE HOSE.

   If the hose has been in place for a long time, it may be stuck. Carefully twist the hose back and forth to wiggle it loose (FIGURE B). If it won't come off, use a utility knife or razor blade to cut it loose. Be aware that excess coolant may run out of the hose as you remove it.

4. INSTALL THE NEW HOSE.

   Put the hose clamps on the hose prior to installing it. As you insert the hose onto the connection, give the hose a twist back and forth to help guide it onto the opening. Some outlets will have a little stop on the end; push the hose up to this point, or push the hose until it sits tight.

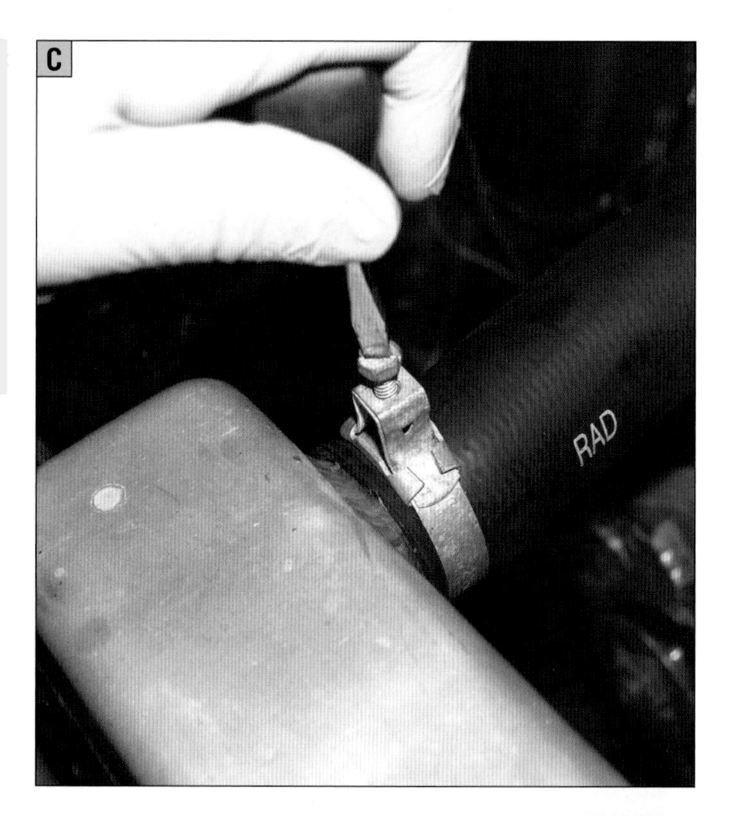

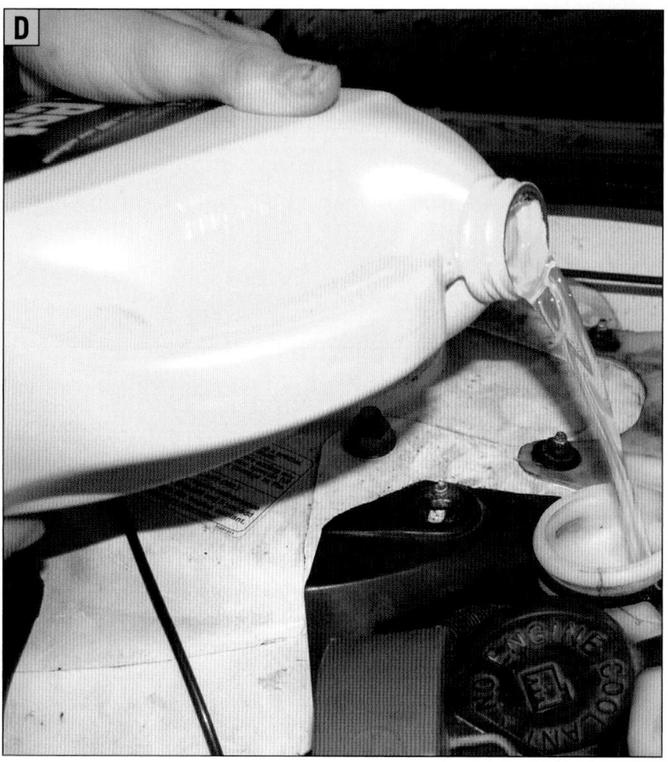

5. **REINSTALL CLAMPS.**
If you have spring-style clamps, simply compress them and move them into place, then release them. Make sure they are not sitting on the inside ridge of the opening, but are past the ridge. This seals the hose against the ridge (FIGURE B).

   If you have compression clamps, use a screwdriver or wrench to tighten them. Be very careful not to overtighten the clamps. The clamp should be tight enough that you can't move the hose, but it should not start digging into the hose itself. The hose should bulge up a little bit from the clamp pressure but not much, and the end should not flare outward.

6. **REFILL THE COOLANT.**
Refill the radiator with the coolant (FIGURE D). Check for any leaks before starting the car. If a clamp connection is leaking, check the clamp placement, and replace the clamp if necessary.

7. **START THE CAR.**
Start the car and bring the engine up to operating temperature. Check for leaks at the connection. If you used compression clamps, double check them after driving for a few days to ensure that they are still tight.

TIPS FOR REPLACING A HOSE

Hose Grease Some mechanics recommend grease, spray lubricants, or oils to make it easier to remove and install the hoses. I prefer to install them dry because some petroleum products can degrade the rubber hoses, silicone sprays will make it easier to get the hose on but harder to take off, and grease can get scraped off inside of the hose outlet and end up in the cooling system. Use care if you decide to use a lubricant to aid in hose installation.

Collapsed Hoses While the cooling system is pressurized, the water pump exerts suction on the lower radiator hose, which can cause it to collapse. Other hoses can collapse, too, but it is most often the lower hose. A hose may collapse for several reasons: it could be worn out, the radiator cap may not be releasing the vacuum properly, or there may be a blockage in the system. If the radiator cap is functioning properly and the system isn't overheating, but the hose has still collapsed, it needs to be replaced.

THE AIR CONDITIONING SYSTEM

IN THIS CHAPTER

How the A/C System Works

The air-conditioning system in your car consists of a condenser, a radiator, an evaporator, a compressor, and an accumulator, as well as a network of hoses, switches, and sensors. It works by taking liquid refrigerant and pushing it through a valve to turn it into a gas, which dramatically lowers the temperature of the refrigerant. In gas form, the refrigerant absorbs the heat from the surrounding air, making the air cold, and the cold air is blown into the vehicle cabin.

Due to the complexity of the A/C system and the special equipment required to service it, all repairs should be done by a qualified technician. However, you should be aware of the basic components of the A/C system and how they work.

Compressor The compressor is a big air pump that is usually run from a belt on the engine. The compressor takes in the refrigerant (usually Freon®) as a vapor and compresses it. The gas heats up as it is compressed, resulting in heat and high pressure vapor. The compressor doesn't run all the time, so the pulley on the front of it usually has a magnetic clutch that can be turned on and off, as needed.

Condenser This piece is usually mounted in front of the radiator. It uses the air that is drawn through the front of the car to lower the temperature of the refrigerant below its boiling point and turn it back into a liquid.

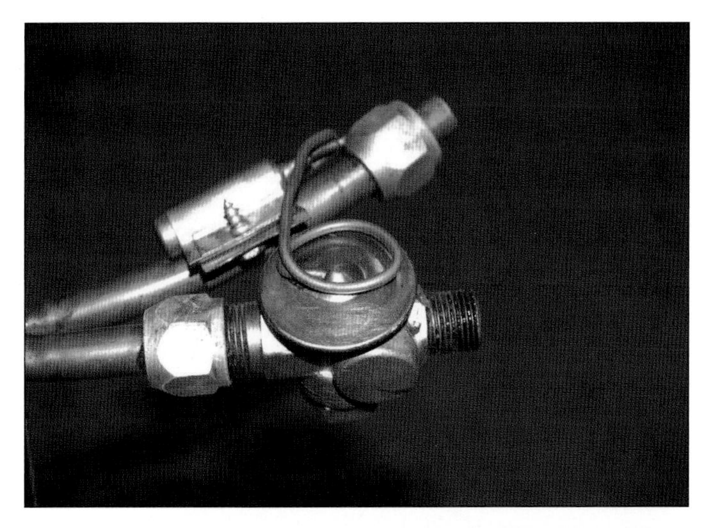

Expansion Valve This valve lets the liquid pass at a lowered pressure and causes something called "flash expansion," which reduces the temperature of the refrigerant and makes it nice and chilly. The refrigerant is mostly liquid with some vapors and is in a state of low pressure and low temperature. Newer systems use an orifice tube, but they perform the same function.

Accumulator Moisture in the A/C system is a problem, so most systems have something to keep the system dry. This may be called an accumulator, a drier, or a receiver. It is usually a cylinder mounted somewhere under the hood and in addition to removing moisture it also acts as a reservoir for the liquid refrigerant when it's not needed.

Evaporator This component is usually mounted under the dash and attached to a fan. As the cooled liquid/vapor mix runs through the evaporator, the fan blows air over the fins, which transfers warmth from the air to the refrigerant, which then cools the cabin air. The refrigerant warms to boiling and changes back into a gas at low pressure. The low pressure gas is then sent back to the compressor and the process is repeated.

Common A/C System Problems

The first indication of a problem with the air conditioner is that it stops blowing cold air. The A/C system is meant to be completely sealed, and when problems arise, it is usually due to gas escaping from the system or contaminants getting into it.

LEAKS

The compressed gases in the A/C system require seals and valves, and over time these will degrade. Pinholes in metal lines can be a problem, too. Keep in mind that a slow leak will eventually turn into a big one. You can add more refrigerant to the A/C system, but the problem will eventually get worse.

COMPRESSOR FAILURE

Major leaks lead to catastrophic problems, like compressor failures. This can happen when the magnetic clutch on the compressor fails, which prevents the compressor from spinning, or the clutch fails to release and the compressor runs all the time.

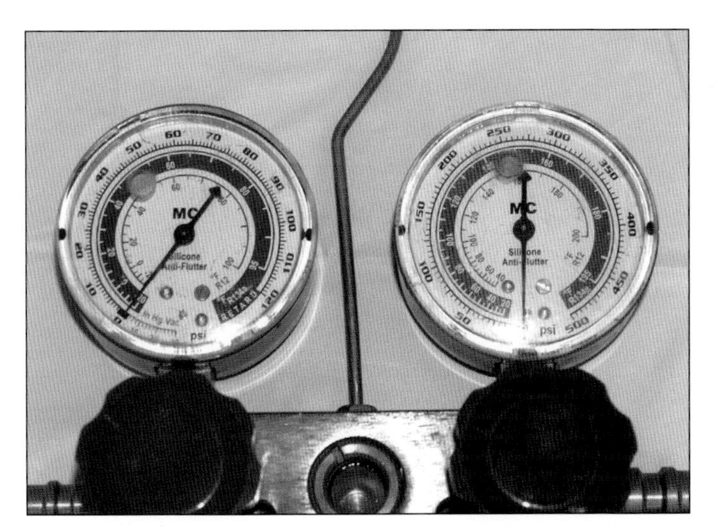

PRESSURE PROBLEMS

Most issues with the A/C system can be diagnosed with the use of a set of gauges. The gauges read the pressure on both the high- and low-pressure sides of the system. Most of the time, checking the low-pressure side is enough to tell you if you need professional assistance.

UNPLEASANT ODOR

Since the air conditioning exchanges cold and hot air, condensation forms on the condenser located under the dash. This drips off and runs out of a drain located at the bottom to the box that holds the condenser. If the drain becomes clogged, it creates a cool, moist area for mold and other unpleasant-smelling things to grow.

CLOGGED FILTER

Newer cars may have a cabin filter to clean the air as it passes through the heater and A/C system. It is usually located under the dash on the passenger side or under a panel in the engine compartment. The filter may become clogged, causing odor or limited performance of the A/C and heater system.

How to: Check for Leaks

Finding leaks in a modern air-conditioning system is difficult. The engine bays are crowded, making it hard to see, and the liquid refrigerant becomes an invisible gas when it escapes. However, there are a few tricks you can use to locate a leak.

WHAT YOU NEED

- Soapy water
- Refrigerant dye
- UV light

First, familiarize yourself the components of your air-conditioning system and locate them in your car. Once you have located one or two of the components, trace all the hoses and lines that connect them and look for leaks using one or more of the following methods.

The "Wet Look" Test The compressor is lubricated with oils, and some of these oils are carried harmlessly through the system. When they come to a leak, the oils get pushed out along with the refrigerant. The refrigerant isn't detectible, but the oils are. Look for an oily "wet look" around a seal or connection that may indicate a leak.

Soapy Water You can also check for leaks by mixing up some soapy water and applying it to the A/C lines and ports with a towel, paint brush, or your fingers. If you get lucky, the leak will start creating bubbles as the gas escapes.

UV Dye Another way to check for leaks is to insert a dye into the system and use a UV light to detect the dye. Some refrigerants come with the dye already in them, and the UV light is a relatively inexpensive purchase. Add the dye to the low pressure side as you would add refrigerant and then run the system, allowing the dye to be pushed through the lines. Shine the UV light on the lines, looking for traces of dye that would indicate a leak.

How to: Add Refrigerant

If your air conditioner starts working erratically or isn't cooling properly, you may be low on refrigerant. This is likely due to a leak, which you should have checked by a professional. However, for a short-term solution, you can recharge the system by "topping off" the refrigerant yourself. Before you begin, be sure to check your local laws. Recharging your A/C system is not allowed in some areas.

WHAT YOU NEED

- Low-pressure refrigerant gauge
- Appropriate refrigerant for your vehicle

A

1. CHECK THE COMPRESSOR. With the car running and the air conditioning set to its maximum output, open the hood and observe the air conditioning compressor clutch (FIGURE A). If the compressor shuts off after a couple of seconds or never spins at all, the system is probably detecting a low refrigerant level.

B

2. DETERMINE THE RIGHT REFRIGERANT FOR YOUR CAR. Look for a label under the hood that shows the type of refrigerant used in your vehicle. This may be on the radiator support or on the underside of the hood (FIGURE B).

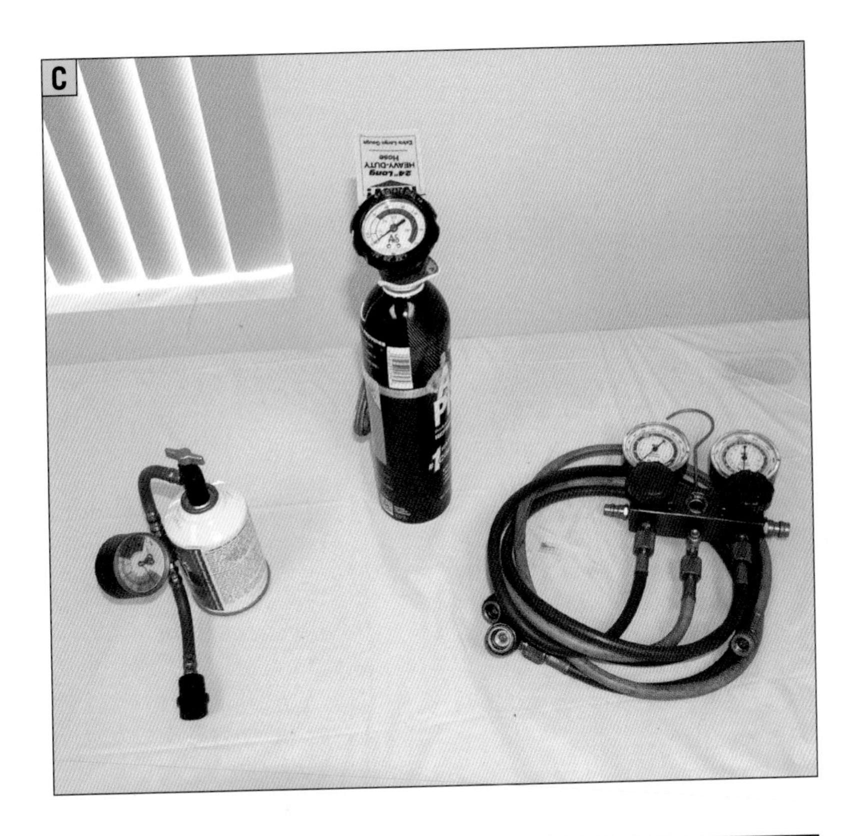

3. GATHER YOUR EQUIPMENT.
   You will need a low-pressure port gauge and refrigerant. Refrigerant is packaged in several ways. Left to right are: a can of refrigerant and a hose with a built-in gauge, a can with the gauge built in, and a set of professional gauges that connect to a can of refrigerant. Any of these will work (FIGURE C).

4. LOCATE THE LOW-PRESSURE PORT.
   There are two ports, or valves, on the A/C system. One is on the high-pressure side and one is on the low-pressure side. The ports should be located near the top of the engine compartment. The low-pressure port will be between the evaporator and the compressor. This one is mounted on the accumulator (FIGURE D).

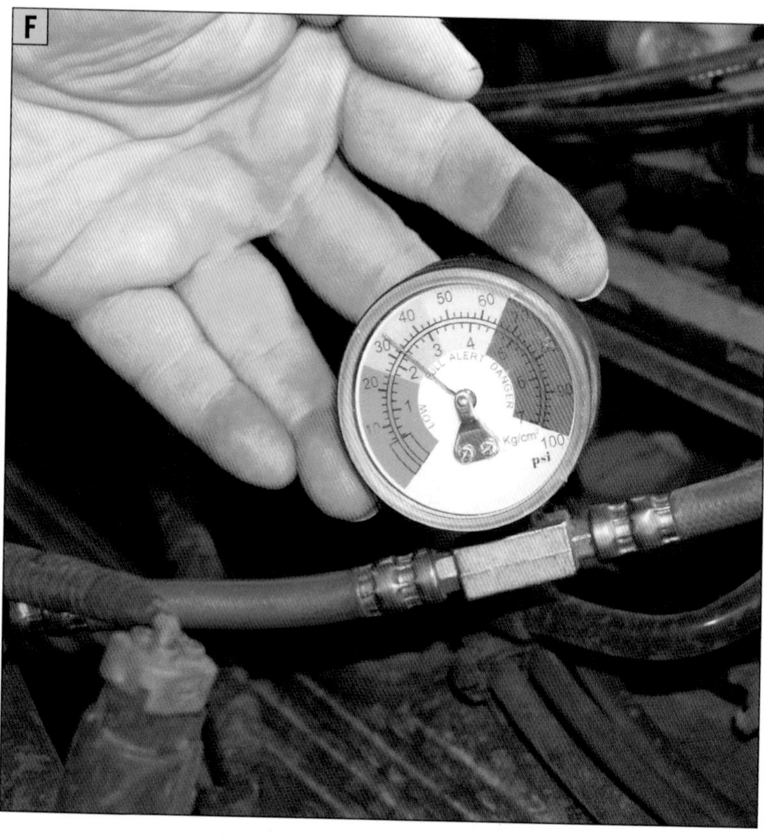

5. CONNECT YOUR GAUGE.
   Twist off the cap from the low-pressure port. Pull the ring on the connection hose back and press the connector onto the port (FIGURE E). You may hear some refrigerant leak—try not to let too much out. Release the ring and the connector should seal against the port.

6. START THE CAR AND TURN ON THE A/C.
   With the car running, set the air conditioning to its maximum output.

7. CHECK THE GAUGE.
   If the pressure is slightly below normal, you can add refrigerant (FIGURE F).

CAUTION

Do not add refrigerant to a full system. Overfilling the system will cause problems. If your gauge is reading extremely low, or doesn't read any pressure at all, do not add refrigerant. You have a major leak that needs to be addressed by a professional.

G

H

8. ADD REFRIGERANT.
   Open the valve on the can of refrigerant and allow it to flow until the compressor is running and the gauge reads "Full" (FIGURE G).

9. REMOVE THE CONNECTOR.
   Pull back the lock ring and pull the connector from the port (FIGURE H). Replace the port cap. Check the pressure after a day or two to see if the system has lost pressure. If it has, you have a significant leak that needs to be addressed by a professional.

How to: Clean a Smelly A/C System

Smells in the A/C system can be caused by two things: mold or a leaking heater core. If the smell is musty and unpleasant, like dirt or rotting leaves, you may have a mold problem. This could be due to a clog in the A/C box drain. If the smell is "sweet" like antifreeze, you may have a heater core leak.

WHAT YOU NEED

- Pressurized air or a blunt instrument to clear the A/C box drain
- Disinfectant spray or mold-killing spray

LOCATE THE SOURCE

Before you can get rid of the smell, you need to figure out what's causing it. Cleaning and deodorizing will take care of mold smells, but a leaking heater core will need to be replaced.

1. CHECK THE CARPET BELOW THE DASH.
   Depending on how the condenser box was designed, heavy leaks can drip down onto the carpet. Since the box is usually on the passenger side, check the carpet by the foot base for leaks from the A/C box. If you find a leak that smells like antifreeze, you will need to have the heater core replaced. If you find a water leak or smell mold, then the drain may be clogged.

2. CHECK THE A/C BOX DRAIN.
   The drain is located at the bottom of the box and drains out under the car, usually in the transmission tunnel or near the middle of the car. You may have to lift the vehicle up safely to view the drain. Look for signs of anti-freeze dripping from the drain. The only thing that should be draining out should be water. If A/C has been running and you don't have condensation dripping out of the drain, it may be clogged.

3. REMOVE OBSTRUCTIONS.
   Professionals use pressurized air to clear a clogged drain. If the drain hole is fairly straight and you don't have access to compressed air, you can try a blunt instrument (like the eraser end of a pencil) to gently move an obstruction. Do not use coat hangers or anything that might puncture the inside of the A/C box.

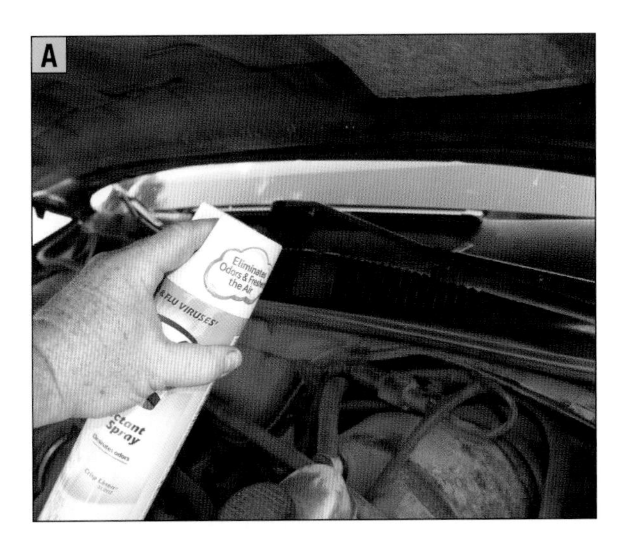

GET RID OF MOLD SMELLS

In order to get rid of a moldy smell, you need to spray a disinfectant onto the condenser to kill the mold growing on it. There are commercial-grade products made specifically for this purpose, or you can use a household disinfectant.

1. **TURN THE CAR ON AND RUN THE A/C AT MAXIMUM.**
   Turn the A/C on full and turn off the recirculating system (usually shown as a picture of your car with air circling around inside). This will allow the A/C to cool with fresh air that's drawn from the outside the car.

2. **SPRAY THE A/C INTAKE WITH DISINFECTANT.**
   With the car running, spray the disinfectant into the intake of the A/C system (FIGURE A). This is usually located just under the windshield. If your car is equipped with a cabin filter (check your owner's manual), remove the filter before spraying the inlet.

3. **SPRAY THE CABIN DUCTS.**
   Turn the A/C system and fan off and spray sanitizer into each of the interior A/C outlets (FIGURE B). Let it sit for a while before re-starting the A/C system.

4. **REPLACE YOUR CABIN FILTER.**
   If you removed your cabin air filter, replace it with a new filter (FIGURE C).

It may take several rounds before the smell goes away. Keep an eye on the drain—if it keeps clogging you will need a professional to remove it and clean it out.

THE EXHAUST SYSTEM

IN THIS CHAPTER

How the Exhaust System Works

With the exception of all-electric vehicles, all cars have an exhaust system to transfer combustible gases away from the engine. The exhaust system also quiets the noise of the engine, reduces emissions to the atmosphere, and helps the engine run more efficiently.

Let's start by identifying some of the more common parts of the exhaust system and what they do.

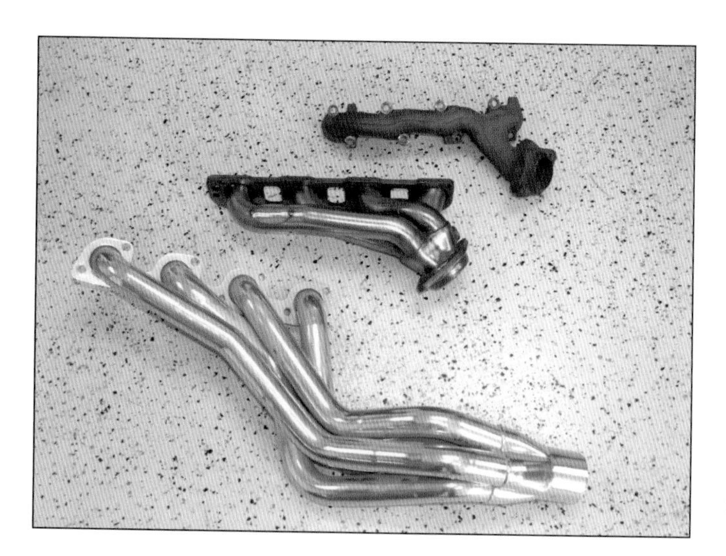

Exhaust Manifold The manifold is typically mounted to the engine and takes the exhaust gases from engine and reduces them down to a single outlet. Manifolds can vary in material and shape. Cast iron manifolds are inexpensive, easy to manufacture, and compact. Larger manifolds made from lighter materials allow the gases to flow more easily than the restrictive cast iron style manifold, but they also take up more space and are harder to install.

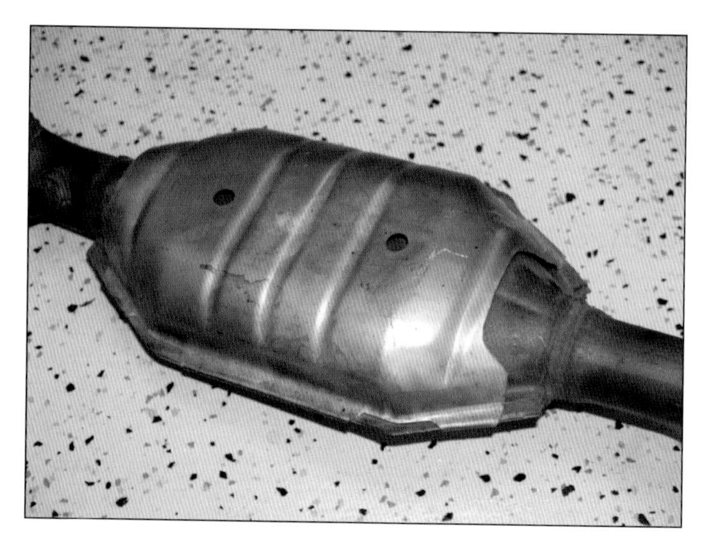

Catalytic Converter The catalytic converter reduces the toxicity of the gases from gasoline and diesel engines.

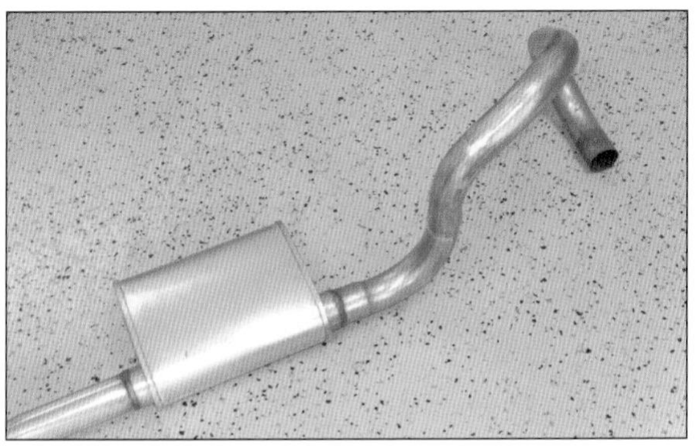

Exhaust Pipe and Muffler System Steel or stainless steel piping is used to transfer the exhaust gases from the engine to the back of the car. The piping may be coated with aluminum or ceramic to help hold in heat and reduce rust.

The muffler quiets the noise of the engine. It may use glass fibers or steel strands to help dissipate the sound. Some mufflers use slots or holes to channel the sound back on itself and quiet the sound.

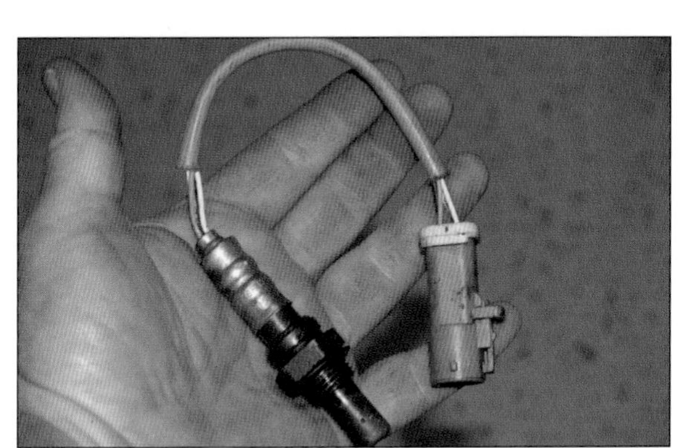

Oxygen Sensor Also called an O_2 sensor, this device measures the amount of oxygen in the exhaust gases and tells the car's computer if the engine is running lean (not enough fuel) or rich (too much fuel). The computer can then adjust the engine fuel system for best efficiency.

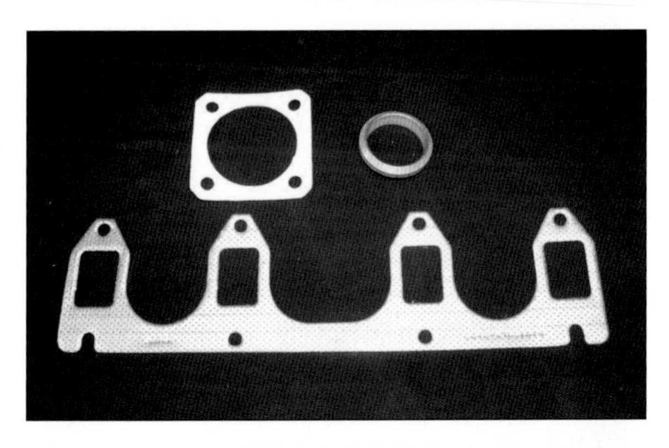

Exhaust Gaskets Gaskets are used at connection points to mate pieces of the exhaust system that may not line up perfectly. They help prevent gases and sounds from escaping, and may be made from graphite, paper, steel, or copper.

Exhaust Hangers The piping is held to the car with flexible holders, which allow the hard metal pieces to move and not crack.

Common Exhaust System Problems

The exhaust system is located under the car, which makes it susceptible to damage from harsh driving conditions. The two biggest problems with the exhaust system are leaks, which can cause loud noise while the engine is running, and emission failures in the catalytic converter or oxygen sensors.

EXHAUST MANIFOLD LEAKS AND WARPING

Exhaust manifolds can crack or warp, and the bolts holding them can loosen, causing an exhaust leak right next to the engine. Depending on the size of the leak, it may sound like a small *tap, tap, tap* or a loud rumble while the engine is running. Leaks at the manifold will usually cause a black stain from the escaping exhaust gas.

GASKET LEAKS

A gasket is a seal that fills the space at the connection point between two components, such as two pipes in the exhaust system. Hot gases can burn away the gaskets and cause leaks. Gasket and pipe leaks result in louder than normal sounds from the exhaust system when the engine is running. Fixing a leaky pipe gasket is something you may be able to do yourself.

CLOGGED OR POISONED CATALYTIC CONVERTER

Excessive oil or carbon can clog a catalytic converter. When this happens, the engine will not accelerate very well and may stall after a few minutes. An engine with a clogged converter may also send an error code from the O_2 sensors to the computer. A clogged converter should be replaced by a professional.

Catalytic converters can also become "poisoned," usually due to an additive run in the engine. These additives coat the converter and cause it to stop working or work less efficiently. Poisoned converters can be detected by the O_2 sensor or by an emissions test.

O_2 SENSOR FAILURE

The oxygen sensors measure the output from the engine and tell the computer how to adjust the fuel and air mixture for most efficient operation. If something is wrong with the O_2 sensor, the "check engine" light will come on, and the computer will post an error code.

EXHAUST LEAKS

Punctured or rusted out exhaust components can cause leaks. Leaks make the engine noise louder, and can be dangerous as they can allow toxic gases into the cabin.

BROKEN HANGERS

The flexible part of the hanger can fail, causing the exhaust system to sag or even break. When a hanger breaks, you may hear a "clunk" as you are driving due to the exhaust system hitting the car, or it can start dragging on the ground. If you find a broken or damaged hanger, have it replaced.

How to: Replace an Exhaust Gasket

Exhaust gaskets can occasionally wear out due to pressure and movement in the exhaust system. While the car is turned off and the system is cold, inspect the gasket areas and pipes for signs of leaks—usually black, sooty debris coming out between the gaskets or out of rust holes in the pipes.

WHAT YOU NEED

- Hand tools matched to the bolts
- Rust-penetrating solvent or spray
- Gasket scraper
- Safety equipment
- Replacement gasket
- Jack and jack stands (if needed)

Gasket surfaces are usually bolted together, so you can do a repair with hand tools. Just remember, the exhaust runs under the car and is subject to road debris, water, and heat. Getting the bolts loose can be very difficult, and sometimes the bolts break.

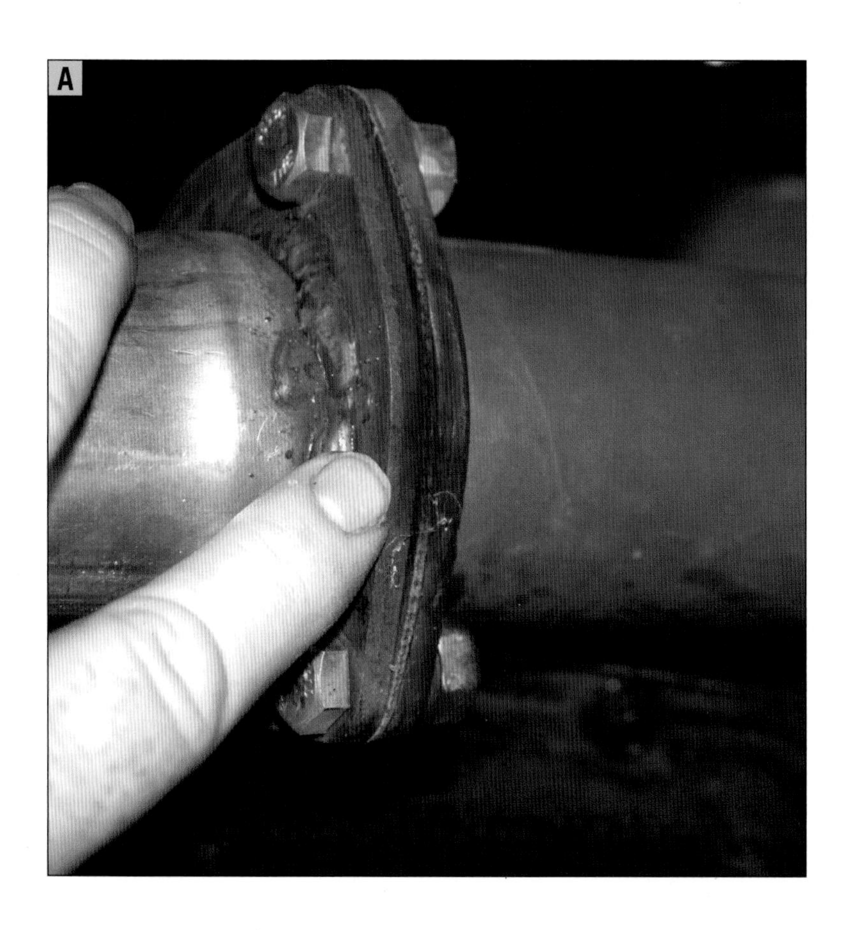
A

1. SAFELY ELEVATE THE CAR. You will likely have to elevate the vehicle in order to access the gaskets. Use a car jack and safety stands to raise the car

2. INSPECT THE GASKET. Check the gasket for signs of a leak. This gasket has been leaking at the header flange. The black powdery residue is evidence of the leak (FIGURE A).

CAUTION

Working on an exhaust system requires equipment that will allow you to be under a car safely. If you need to raise the car, make sure you can do so safely. Always work on the exhaust system while it is cold, especially when doing repairs near the catalytic converter.

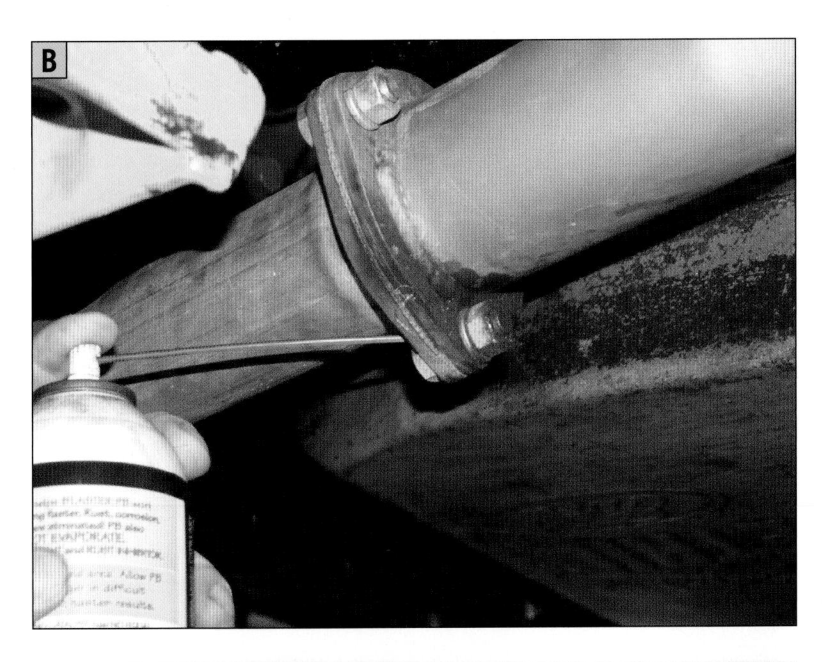

3. LOOSEN THE BOLTS.
   Spray rust-penetrating solvent on the bolts and loosen them with the appropriate tool. You may need more than one application to loosen the bolts (FIGURE B).

4. SUPPORT THE EXHAUST SYSTEM.
   Depending on where the exhaust system is connected to the car, you may have to support the exhaust system after you loosen the bolts. You can use an extra jack stand, as shown (FIGURE C).

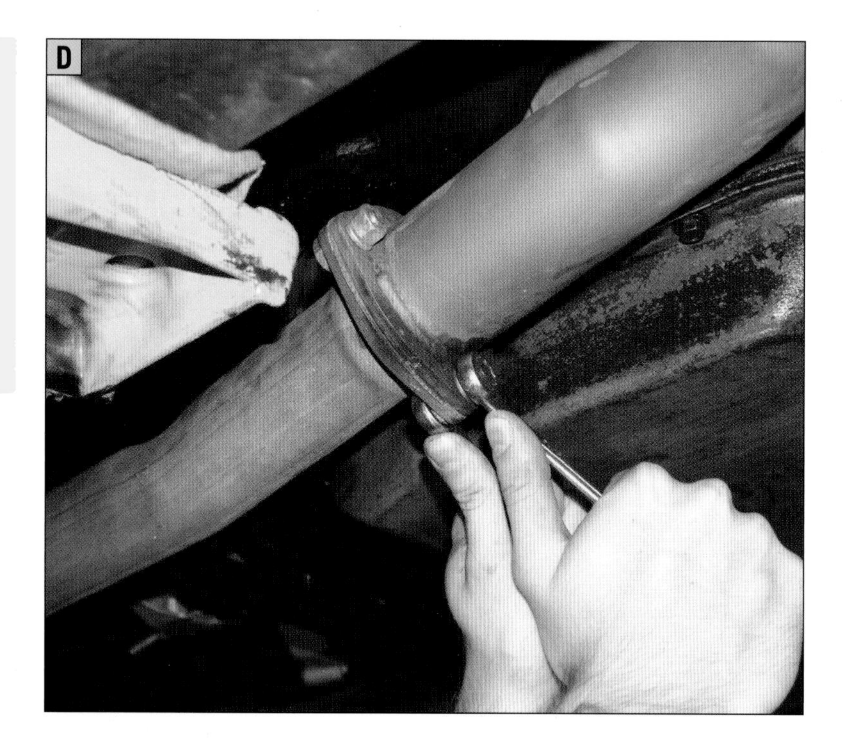

D

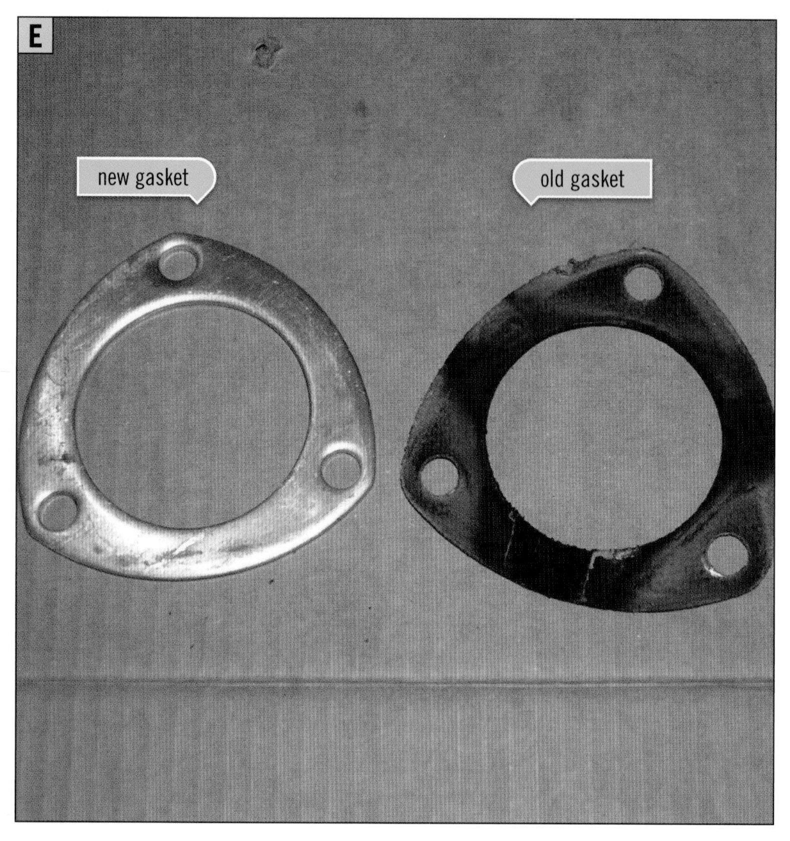

5. **REMOVE THE BOLTS.**
   Using the correct size wrenches or sockets, remove the hardware from the exhaust (FIGURE D).

6. **INSPECT YOUR OLD GASKET.**
   Check the old gasket to see what failed. This old gasket shows both a breakdown of the material (bottom of the gasket) and black exhaust powder escaping between the bolts, an indication that the bolts weren't tight enough at this connection (FIGURE E). A thicker set of gaskets made of soft aluminum will help with the sealing.

   If the bolts or hardware are getting rusty or the threads are getting worn, replace them with correct strength hardware.

F

G

7. **INSPECT THE MATING SURFACES.**
   Damage or warping may have caused the gasket to fail. If it has, you may have to have that part of the exhaust system replaced, as the gasket will probably fail again. Scrape off any old gasket with a gasket scraper before installing a new one (FIGURE F).

8. **INSTALL THE NEW GASKET.**
   Move the pipes into position by hand. Install the new gasket between the pipes and tighten the bolts so the gasket seals completely (FIGURE G). Take the car off the safety stands, start the engine, and listen for any leaks from the new gasket. After a day or two, tighten the bolts again—the heat may loosen them.

How to: Change an Oxygen Sensor

Oxygen sensors analyze the oxygen levels in the exhaust gases from a combustion engine. They tell the car's computer if it needs to add or take away fuel to make it run more efficiently. They also measure the efficiency of the catalytic converter. If you are getting consistent error codes from the computer, the sensor probably needs to be changed.

WHAT YOU NEED

- Computer code reader
- Sensor socket and wrench, or similar tool
- Protective gloves
- Jack and jack stands (if needed)
- Anti-seize compound
- Tools to unhook and reconnect the battery

A

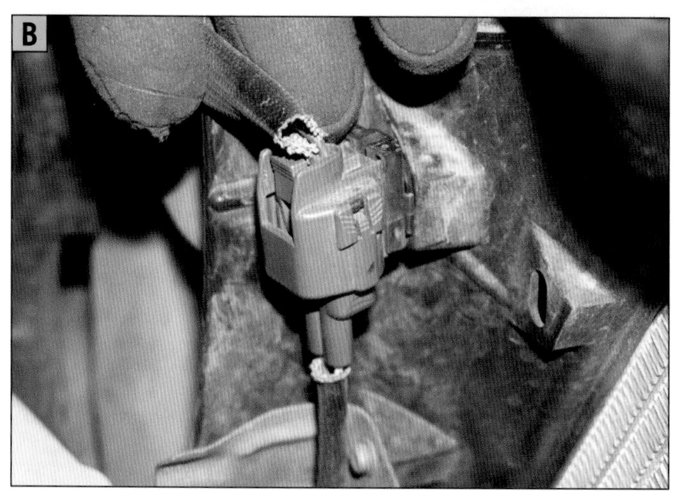

B

1. LOCATE THE SENSOR.
With the engine cold, locate the oxygen sensors on the exhaust system. There are sensors in two locations: between the exhaust manifold and the catalytic converter, and after the catalytic converter. Each location may have one or two sensors. In some cases, you may need to raise the vehicle to access the sensors. If this is the case, elevate the car safely using jack stands (FIGURE A).

2. CHECK THE CONNECTOR.
Before removing the sensors, check the connections to the sensors. They are frequently frayed or burnt, and may be the cause of the sensor not working. This connector is in good shape (FIGURE B).

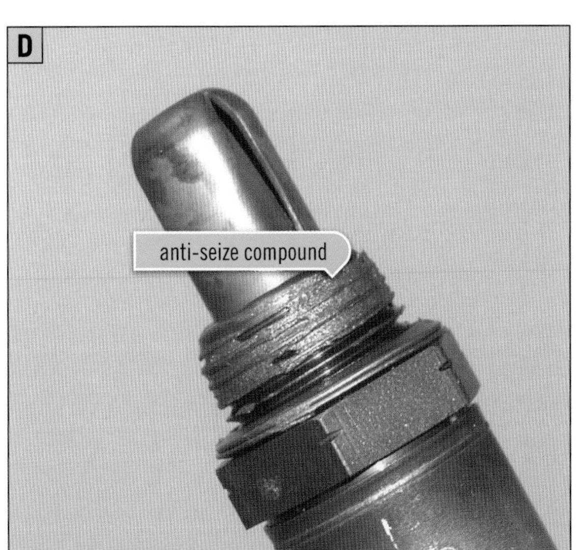

3. REMOVE THE SENSOR.
   Unhook the negative battery cable and then unhook the sensor wires before removing the sensor (FIGURE C). The sensor unscrews like a spark plug or a bolt. Be careful to not strip the threads of the exhaust pipe when removing the sensor. Sensors can be difficult to remove cold, so try running the engine for a minute or two to warm up the pipes or manifold without getting it too hot. If the engine is warm, be sure to wear protective gloves and avoid burns.

4. APPLY ANTI-SEIZE COMPOUND IF NEEDED.
   Most new sensors come with anti-seize compound already applied. Look for a silver-colored substance on the threads (FIGURE D). If your sensor doesn't have it, apply some to the threads to make it easier to remove in the future.

5. INSTALL THE NEW SENSOR, RECONNECT THE SENSOR WIRES, AND CLEAR ERROR CODES.
   Tighten the sensor to the manufacturer's specification, reconnect the sensor wires and the battery, and erase the codes. Bring the engine up to operating temperature and see if the codes return. If they do, you will need to have a professional check the car.

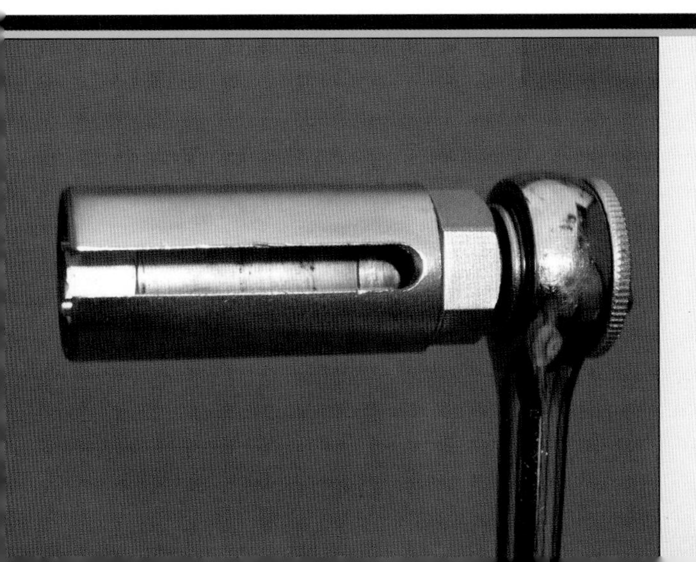

SENSOR SOCKET AND WRENCH

Sometimes you can use a regular wrench to remove a sensor, but if you need a socket to access it, you will need a socket with an access slot for the wires to come through.

THE COMPUTER SYSTEM

IN THIS CHAPTER

What the Computer System Does

Cars first began using computers in the early 1980s. Initially, computers only controlled the ignition system, but over time they have taken over more and more operations. They now control the power to different devices, shift the transmission, and can parallel park and watch for other vehicles.

All of that control does come with a price. Because the newer cars rely on the computers for better mileage and emissions, computers are now an active part of the system. The newer your car, the more likely you are to be stranded if the computer fails. These are the main components that contribute to the computer system.

Computer The on-board computer in your car may be referred to as an ECU, ECM, or PCM. Computers are generally located in a place away from heat and weather, so on some models they can be difficult to access. The computer communicates with other systems, such as the anti-lock brakes and tire monitors, and makes constant adjustments to your engine while you drive. It records any problems that occur while you're driving, and can tell a mechanic exactly what you were doing when the problem occurred (and that includes speeding!).

Computer Port The computer "talks" to the rest of the vehicle through a port called the OCD or OBDII port. It communicates using codes that can be read using a code reader. In the United States, cars use a set of codes called OBD (on-board diagnostic). European vehicles use EOBD (European On Board Diagnostics). The OBD port is usually found underneath the dash on the driver's side and has a distinct "D" shape.

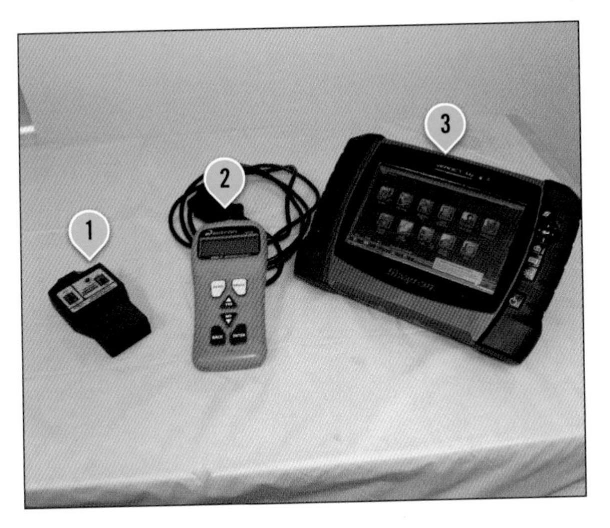

Code Reader A code reader is used to retrieve the error codes on most newer vehicles. The code reader on the left (1) is for older Ford vehicles with OBDI. The one in the middle (2) is for reading OBDII pre–2003, and it has the D-shaped connector. The one on the right (3) is for newer cars and has the same D-shaped plug, but will read the newer, faster computer communications.

You should know how to read the codes on your car, and if you plan to keep your car for a while, you should consider purchasing a code reader. Make sure you get the right reader for your vehicle.

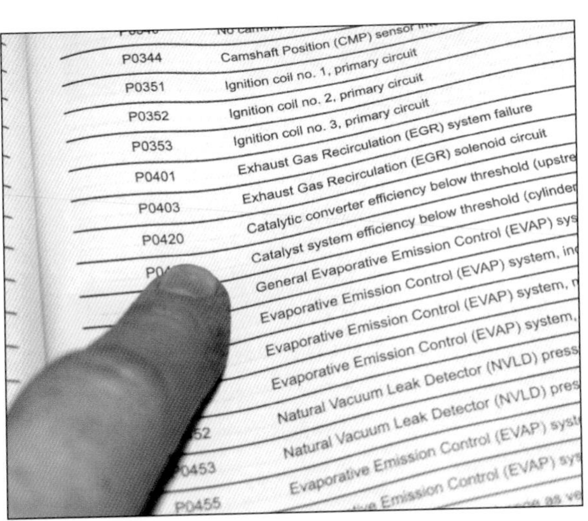

Computer Codes To figure out what the codes on a code reader stand for, you'll need to consult your vehicle shop manual or go online. Depending on the year of your car, it may use three-digit or four-digit codes. OBDII codes begin with a "P" followed by a four-digit number. These are also referred to as DTC codes (diagnostic trouble codes). Some codes are generic and some are specific to the manufacturer.

"Check-Engine" Light The "check engine" light on the dash is part of the on-board computer system. If the computer finds a problem that it cannot correct, it will illuminate the "check engine" light to let you know there is a problem. If this happens, don't panic; it may not be a serious problem , but it should be investigated.

Common Computer System Problems

Troubleshooting the computer is not difficult. The computer itself will report the problem, or your car will stop running or run very poorly. The following issues may indicate an issue with the computer.

"Check engine" light does not come on at start-up. When you turn the key without starting the car, your "check engine" light should come on. This is the computer's way of telling you it is ready to go. If you don't see the "check engine" light flash when you turn the key, the computer may detect a problem.

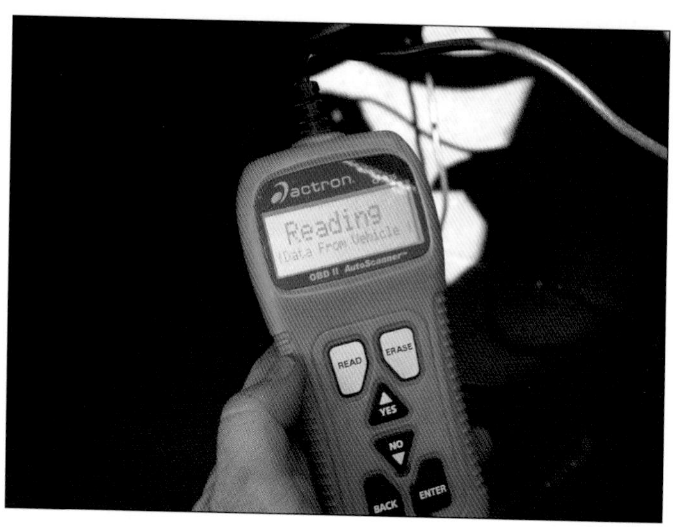

The code reader does not come on. When you plug your reader in, it should power up. If it doesn't, you may have a problem with the power to the computer, or the computer system itself.

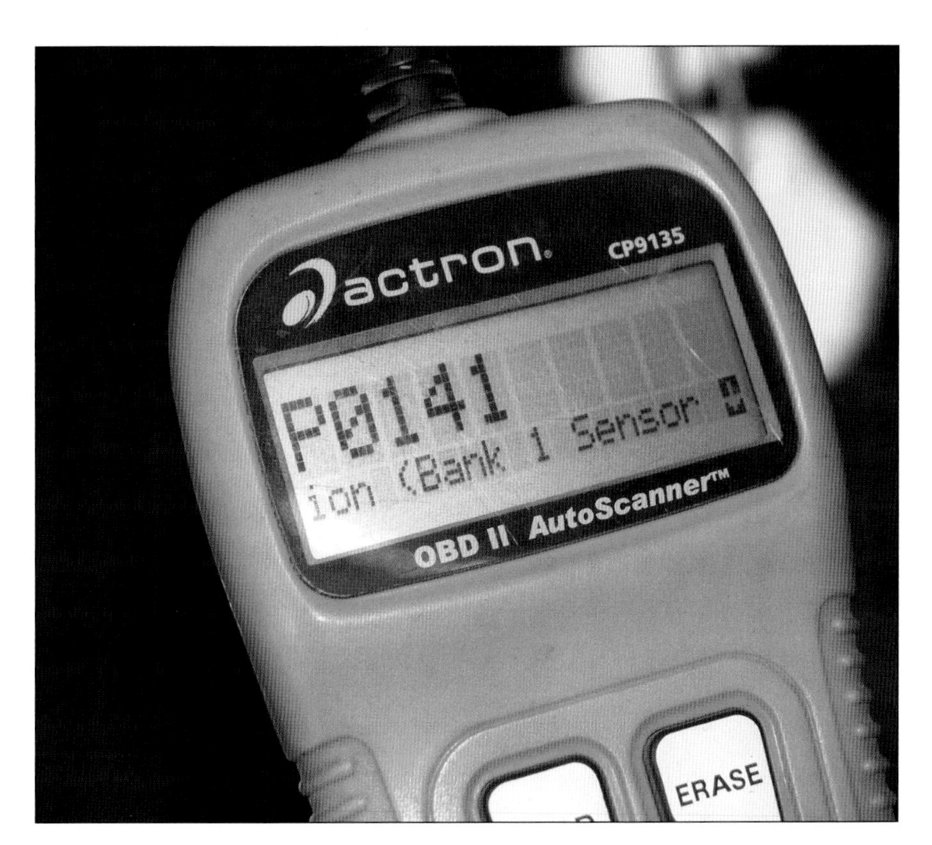

The code reader has an error. If you have the correct code reader for your car and you cannot communicate with the computer, then there may be a problem with the computer. You will need to take it to a professional.

KEYLESS IGNITION AND PUSH-TO-START SYSTEMS

Newer cars equipped with push-to-start ignition do not have a "run" position like older cars. These cars will usually indicate they are ready by flashing a display on the information screen or instrument panel immediately after you push the start button. This display varies by manufacturer; consult your owner's manual for the correct signal that your car is ready to run.

How to: Check for Codes

If your car's computer detects a problem, it will record an error code and activate the "check engine" light. You can check the computer code using a code reader. Although some newer vehicles with on-board screens may be able to display the error codes without the use of a reader. Consult your owner's manual for more information.

WHAT YOU NEED

- Computer code reader appropriate for your vehicle

If you need to purchase a code reader, check with an auto parts retailer to determine which model is correct for your vehicle.

A

1. LOCATE THE COMPUTER PORT.
   Locate the D-shaped connector on the car. It is usually under the driver's side dashboard (FIGURE A).

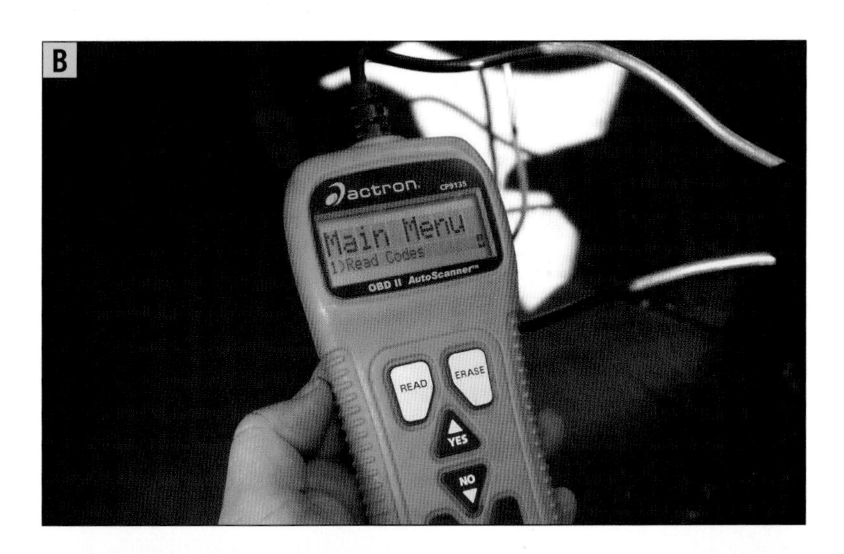

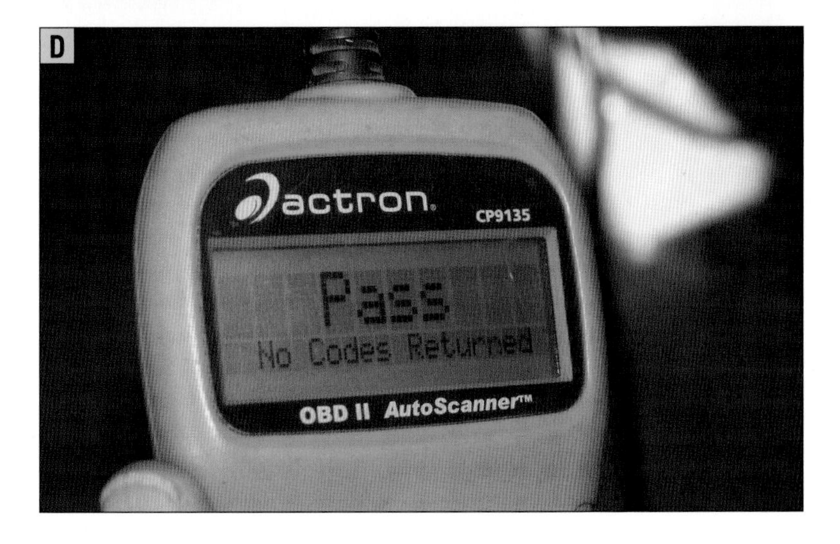

2. PLUG IN THE CODE READER.
   When you plug in your reader, it should power up and be ready to read codes without the car turned on (FIGURE B).

3. TURN THE CAR ON, BUT DON'T START THE ENGINE.
   Turn your key to "on," but do not start the vehicle. If your reader prompts you to "read codes," push "yes." On cars with push-to-start ignitions, it may work right away (FIGURE C). Check your owner's manual for more information.

4. CHECK THE READER.
   If no codes are found, it will report "no codes" (FIGURE D). In that case, turn off your car and unplug your reader. If your computer has stored error codes, it will tell you "x of x" codes found.

E

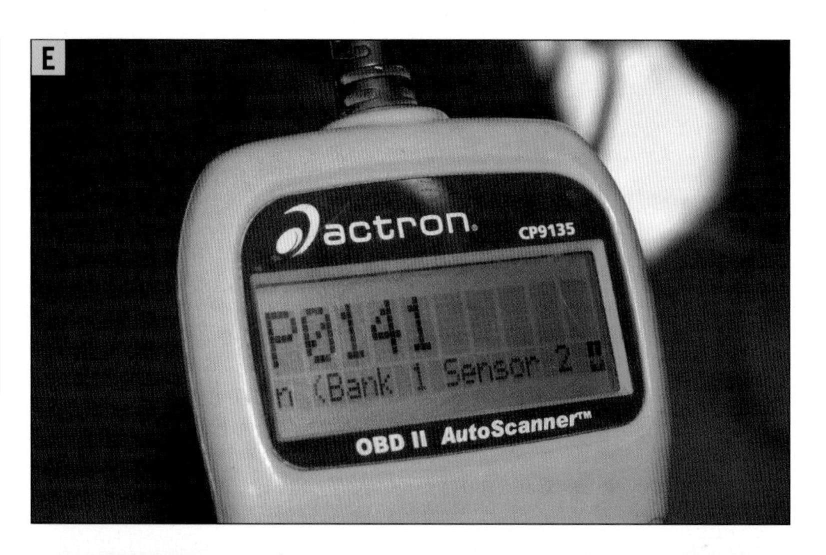

F

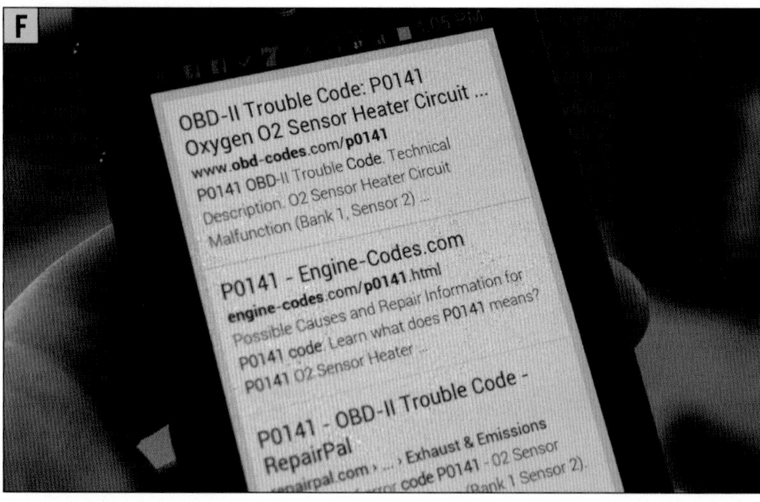

G

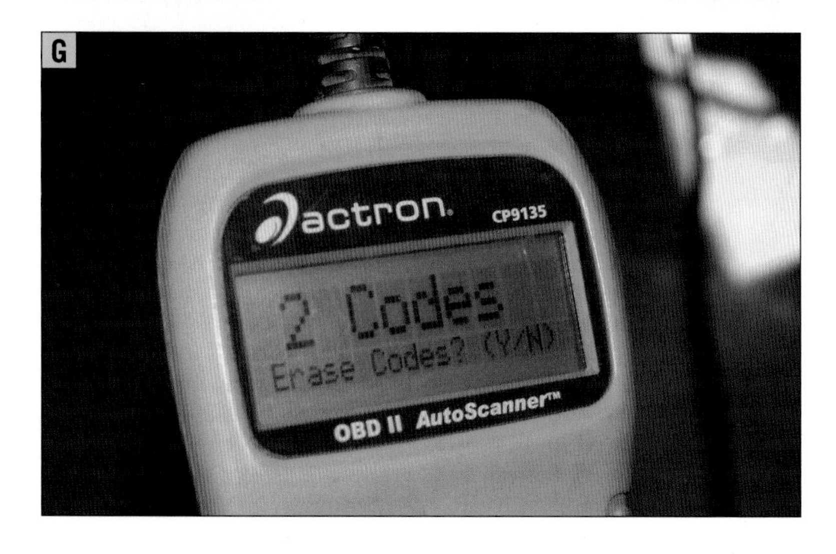

5. TAKE NOTE OF ERROR CODES.
   If you have error codes, the reader will display a "P" followed by the four-digit code (FIGURE E). Take note of these codes.

6. DETERMINE WHAT THE CODES MEAN.
   Your reader may tell you what the code means, or you can look up the number in your shop manual or online (FIGURE F). Depending on how sophisticated your reader is, it may tell you in more detail what your code actually means. If it isn't something you can fix or understand, write down the codes to share with a repair professional.

7. ERASE CODES. (OPTIONAL)
   Error codes are stored in the computer until they are erased. They are not erased when the problem is solved, but if you solve the problem, you can erase the codes so they don't cause confusion later (FIGURE G). If you are unsure about the code, don't erase it—leave it for a professional to check.

WHEN THE "CHECK ENGINE" LIGHT TURNS ON

You're out driving and that little light just came on. What should you do? First of all, don't panic. The light can come on for hundreds of reasons, and it doesn't always mean your engine is about to go.

Check your other gauges and lights. If they are all off and the vehicle is running and sounding normal, you don't need to pull over right away. Check oil pressure, engine temperature, tire pressure, and any other gauges on your car. If one of these lights is on, pull over when it is safe to do so and inspect the car. If the car is running poorly and the "check engine" light is on, pull over to a safe location as quickly as possible.

If you have a code reader available, perform a computer code check. Although the repair shop will read it themselves, it is good to have the same code information they have.

THE ELECTRICAL SYSTEM

IN THIS CHAPTER

How the Electrical System Works

Common Electrical System Problems

Troubleshooting the Starting System

How to: Replace a Taillight Bulb

How to: Replace a Headlight Bulb

How to: Check the Charging System

How to: Clean the Battery Terminals

How to: Change a Battery

How the Electrical System Works

Almost everything on a modern car relies on electricity. This electricity comes from both the alternator, which generates energy, and the battery, which stores energy generated by the alternator. The amount of electricity your car is using at any time is called the *load*. How much load is on the system will determine whether the car draws power from the alternator, the battery, or both.

The following are some of the primary components of the electrical system.

Battery Your car has one or more batteries to store electricity. The starting system requires a lot of power, which it draws from the battery. Once the engine starts, the alternator begins recharging the battery.

Electric vehicles use multiple batteries, which are designed to be used until most of the power they contain has been drained. The batteries turn the electric motor, so there is no "starting" the engine.

Alternator The alternator generates power while the engine is running. It is usually driven by a belt on the front of the engine. The power generated by the alternator is used for both charging the car's battery and running the electrical components of the car. When many electrical components are in use and the engine is running slowly, like at idle, the battery may help run the car.

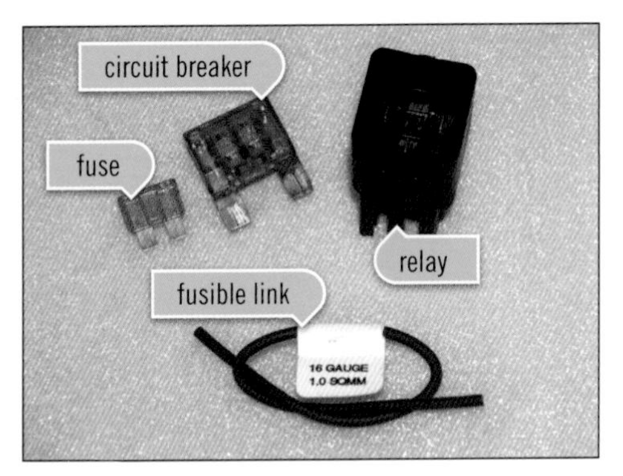

Fuses, Circuit Breakers, and Relays Fuses and circuit breakers are designed protect the electrical components by stopping the flow of electricity if it gets too high. Relays are used to turn on and off power using a secondary power source. The starter solenoid is a big relay that uses power from the key switch to turn on the much larger voltage to run the starter. Fusible links look like wires and will melt to stop the flow of power, like a fuse.

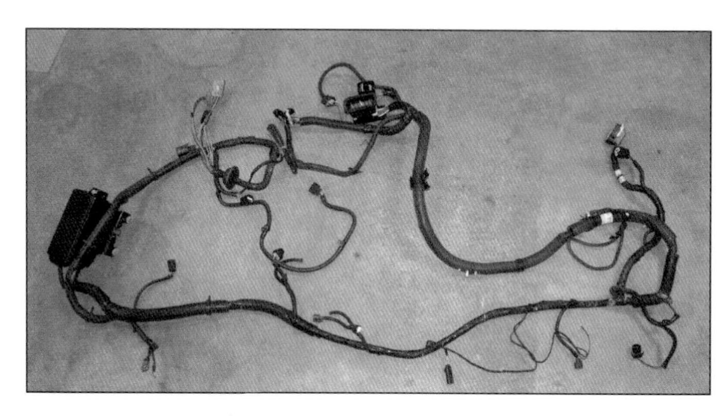

Wiring Electricity is carried through wiring. The thicker the wire, the more power it can carry.

Lighting The light bulbs in your car may be incandescent, LED (for reliability), or xenon (for greater light output).

Sensors Sensors take measurements from the drivetrain and communicate these measurements to the computer, which then adjusts the fuel-to-air ratio to keep the car running efficiently.

Engine Starter The engine uses an electric starter to turn over the motor. When the starter is turned on, it spins a wheel called a *flywheel,* or *flexplate,* to start the engine.

Common Electrical System Problems

Electrical problems can be some of the most frustrating issues to diagnose. Because there are so many components in your car that depend on electricity, many different problems can occur, and a problem with one component can affect others.

With the exception of individual components failing (which they do, frequently), here are some of the most common electrical problems.

BLOWN FUSES AND CIRCUIT BREAKERS

Fuses are used to prevent damage to the electrical circuit. They sacrifice themselves by "blowing" if too much power tries to surge to the electrical components.

Learn how to change a blown fuse, and keep a small supply of extra fuses handy in your car. Check your owner's manual for the location of the fuse box and a list of what each fuse safeguards.

BURNED-OUT BULBS

Exterior bulbs usually burn out first because they are used most frequently. Your ability to change bulbs depends on the design of your car, but in general, bulbs located deep within the car are more difficult to change

In many newer cars, LED lights are used instead of incandescent bulbs because they last a very long time. Consider switching to LEDs, if you can.

DEAD BATTERY

Automotive batteries have a finite lifespan and need to be changed regularly. You may have heard that cold weather will shorten the life of a battery, but it's actually hot temperatures that cause damage. Heat will expand the fluids in the battery, shortening its life. Overcharging the battery can also cause damage. Damage to batteries usually occurs in the summer, and the failure happens in the winter.

When changing a battery, make sure your new battery is the proper size and that the rating in cranking amps is at least as high as the original battery.

CHARGING SYSTEM FAILURE

The charging system runs off the engine and creates power to recharge the battery and run the electrical components while the car is operating. The battery relies on the charging system to keep it topped off with power, and when it fails, the battery is the only power source keeping the engine running and the lights on. The charging system can fail by not charging enough (or at all), or by overcharging and damaging the battery. Keeping the battery connections clean is essential for the charging system to work properly.

Also, if you customize your car with power-hungry electrical components, such as big stereos and off-road lights, make sure your charging system is capable of covering the added energy needs.

STARTING SYSTEM FAILURE

The starting system uses more power than anything else on your car. The starter can wear out due to heat from the exhaust, corrosion, wear on the teeth that turn the motor, or the battery can simply run out of power.

Because the starting system uses so much power, it is important that the wiring is capable of carrying the needed power, and that the contacts are kept clean. Inspect the wiring frequently for damage and corrosion, and to ensure that they make a good connection.

GROUNDING ISSUES

An electrical circuit must travel back to where it started (the battery) to operate properly. To do this, your car uses the metal parts of the chassis as a grounding source for the electrical circuits, and the ground wire is connected to the battery.

Grounding issues are frequently the problem when you have a faulty circuit. The bolts holding the ground wires can loosen, and the wires can be damaged or become corroded.

Some ground wires are buried deep in the car and should be left to the pros to find and repair. If you notice the ground points while working on your car, make sure that the contact to the metal frame is good and not corroded. Try to keep them clean.

Troubleshooting the Starting System

The starting system in most cars consists of four basic elements: the starter, the starter solenoid, the battery, and the ignition key. When you turn the ignition key to start, it sends a voltage to the starter solenoid, which closes a connection and allows the battery to send a lot of power to the starter to turn over the motor.

Problems with the starting system may lead to the following issues:

WHEN YOU TRY TO START THE CAR, ALL THE POWER SEEMS TO GO AWAY.

When you turn the key, the lights go out and the car seems to lose all power. It may or may not come right back.

This is usually an indication of a battery that is drained or is not taking a charge. Check all the connections to the starter for corrosion, as this can cause the wires to not pass the power along. Check the battery for proper voltage.

NOTHING HAPPENS WHEN YOU TURN THE KEY.

If you turn the key and the engine doesn't start, but the lights stay on and the battery is charged, it's likely that there is a problem with the solenoid. Either it is not getting power from the ignition switch, or it has gone bad. In rare instances, this problem can be caused by a bad ignition switch.

Check the connections to the solenoid for corrosion and tight fit. The solenoid may be located on the starter, in the engine compartment, or in the fuse box. Don't try to bypass the solenoid by running power directly to the starter from the battery; this can be dangerous.

Some cars have a fuse protecting the power to the starter solenoid. If this fuse is blown, the solenoid won't function. Check your owner's manual for the location of this fuse and replace it.

YOU TURN THE KEY AND HEAR A CLICK, BUT THE MOTOR DOESN'T TURN.

The "click" you hear is the internal switch in the solenoid trying to make contact. It is either not able to make contact because of wear on the inside, or it made contact and the starter is not working.

Check the power from your battery and make sure it isn't low, and look for loose or corroded connections. Depending on the location of the solenoid, you may be able to change it.

A hard click usually means the solenoid is working, but the starter is not. If the solenoid is accessible, you can check the output from the solenoid with a voltage meter to see if the power is being sent to the starter.

YOU TURN THE KEY AND HEAR A SERIES OF RAPID CLICKS.

This is another sign of low voltage. The solenoid is closing, but there isn't enough power to keep it closed, so the switch opens up again. This causes the battery to recover and send power to the ignition switch, which makes a rapid clicking sound. If you hear this, charge the battery and check the contacts.

THE ENGINE TURNS OVER, BUT IT IS SLOW AND WON'T START.

This is a sign of low power or a bad starter. If it happens more with heat, it's a sign the starter is going bad. If it happens while cold, it could be the battery or the starter. Check the power output from the battery first.

Another possible cause is a problem with the engine. Something may have seized in the engine, the wrong oil may have been used, or the timing may be incorrect.

THE STARTING MOTOR IS SPINNING, BUT THE ENGINE DOESN'T TURN OVER.

If you hear a high-speed whirring from the engine, it means something is broken in the starter or that the starter is loose and not engaging the engine. You can check the starter to see if it is loose, but you will likely need to replace it.

How to: Replace a Taillight Bulb

Changing light bulbs on your car can be easy or difficult, depending on where they are located. Taillights and interior lights are usually easy to access, but other bulbs, such as dashboard lights, require major disassembly. Before you begin, assess your ability to locate and change the bulb.

WHAT YOU NEED

- Replacement bulbs
- Gloves or a clean cloth
- Tools to remove guards and panels (if needed)
- Dielectric grease or petroleum jelly (optional)

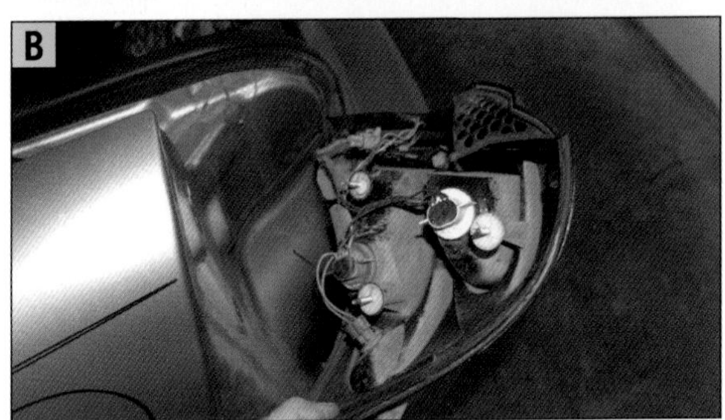

1. REMOVE ANY GUARDS OR PANELS.
   You may need to remove a panel or guard to get to the taillights. To access the taillights on this car, we needed to remove this panel, which is located in the trunk of the car (FIGURE A).

2. REMOVE THE BULB.
   On some vehicles, like this one, the entire taillight assembly comes loose after removing a few screws. On other cars, you can access the connectors without removing the assembly (FIGURE B).

   The bulbs usually require a gentle twist to pull them from the socket. Remove the bulb from the taillight carefully so you do not break the bulb.

3. INSPECT THE BULB AND SOCKET.
   With the bulb out, inspect the bulb and socket seal for damage and check the socket for corrosion (FIGURE C). If it is corroded, use an electrical contact cleaner and a cotton swab to clean the contact.

CAUTION

Although they don't put out as much heat as a halogen headlight bulb, it's a good idea to wear gloves or use a clean rag while handling any light bulb to prevent oils and dirt from coming in contact with them.

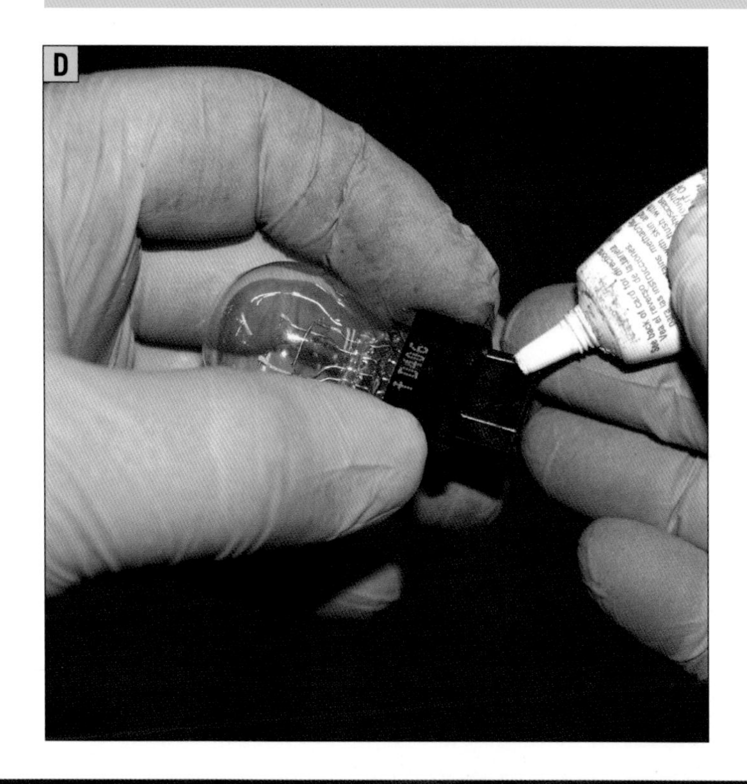
D

4. APPLY GREASE TO BULB.
   A little dielectric grease or petroleum jelly will prevent corrosion from building up on the electrical contacts (FIGURE D).

5. INSTALL THE NEW BULB.
   Installation is the reverse of removal. Use caution when installing the bulb—it should slide or twist into place smoothly.

6. TEST THE BULB.
   Turn on the vehicle and test the new bulb operation. If it doesn't light, remove the bulb and inspect it for damage. If the bulb is okay, you may have a problem with the wiring leading to the socket.

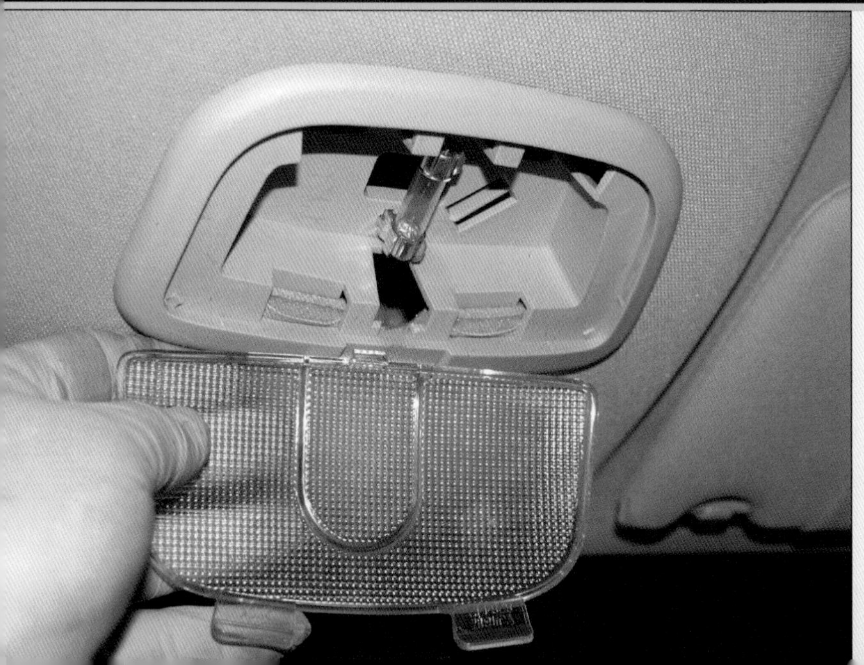

CHANGING INTERIOR BULBS

Most interior bulbs are easy to change and usually don't require tools. This dome light cover unsnapped and we were able to pull the bulb by hand. Some cars may require a screwdriver or simple tools to change the interior bulbs. Be careful touching the glass portion, use a little grease if you have it, and be gentle with the bulbs.

How to: Replace a Headlight Bulb

Changing a headlight bulb is different from changing some of the other bulbs in your car. Because of the high output, they generate more heat than regular bulbs and require extra caution. Avoid touching the glass part of the bulb, and wear gloves or use a cloth when handling the bulbs. Oils from your fingers can be transferred to the bulb, which can cause the glass to heat up and result in the bulb burning out prematurely.

WHAT YOU NEED

- Replacement bulb
- Gloves and safety glasses
- Tools to remove brackets or covers (if needed)

This tutorial shows how to change a halogen bulb, which is the most common style of headlight bulb. Always replace bulbs in pairs.

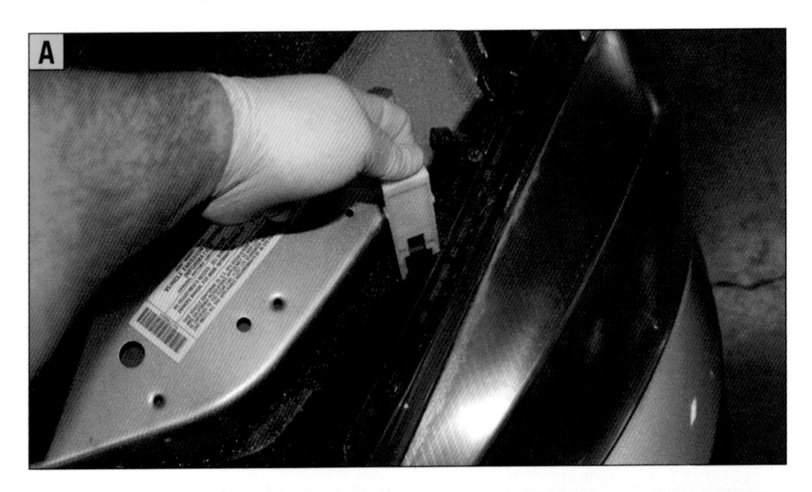
A

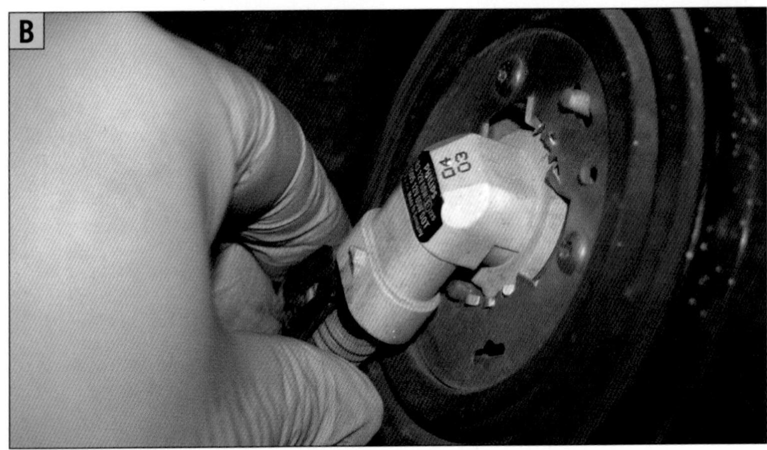

B

1. ACCESS THE HEADLIGHT BULB.
Accessing the headlight can be a challenge on some cars. You may have to remove several covers or brackets to access the bulb, or you may have to snake your arm down between parts to get to it. The bulb on this car is accessed by pulling on a plastic bracket, which allows the headlight assembly to snap out (FIGURE A).

2. DISCONNECT THE BULB.
On some cars, the socket stays attached to the wires. On other cars, like this one, the bulb has a connector (FIGURE B). Gently pull the locking tabs back and pull the connector out of the bulb.

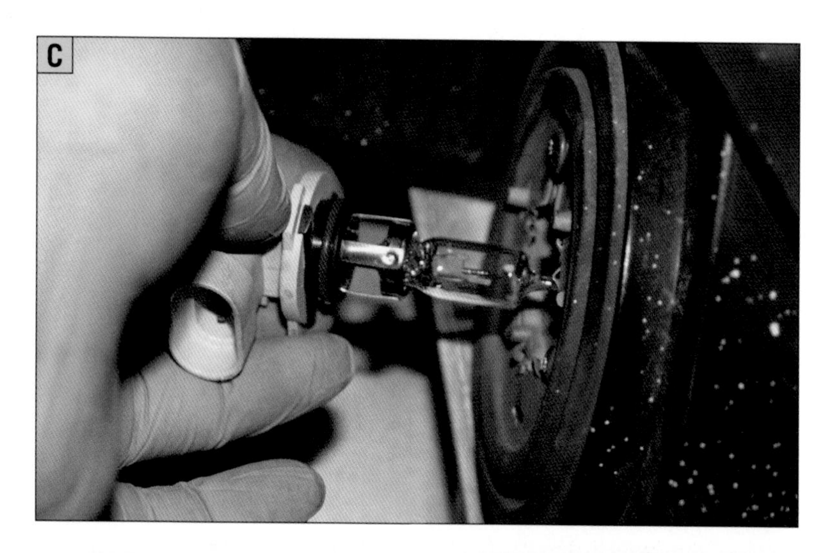
C

D

E

3. REMOVE THE BULB.
Some bulbs will have clips or brackets holding them in place; remove these brackets first. Some cars have a plastic retaining ring that twists and stays on the wiring harness. Be careful not to break the glass when removing the bulb (FIGURE C).

4. INSPECT THE BULB.
Check the bulb for broken filaments (FIGURE D).

5. INSTALL THE NEW BULB.
Installing the new bulb is the reverse of removing it. There is usually a tab or notch on the housing or bulb to prevent the bulb from being installed incorrectly (FIGURE E). After inserting the new bulb, reinstall any brackets or clips that were holding the bulb in place and put back any items you needed to remove to access the headlight.

6. TEST THE HEADLIGHT.
If the new bulb does not light, pull it and check for damage. If it is fine, inspect the wires going to the bulb for damage.

How to: Check the Charging System

The charging system not only keeps the battery charged, but once the engine is running, it provides power to the rest of the car. If the charging system fails, your car will start using battery power to run, but it won't last for very long. You can check the system with a voltmeter.

WHAT YOU NEED

- Voltmeter
- Gloves and safety glasses

READING CAR GAUGES

Most cars have a voltage meter, ammeter gauge, or a warning light. The **voltage meter** measures the voltage of the battery and charging system. When the key is on but the car isn't running, the voltage meter shows the voltage of the battery. When the car is running, it shows the voltage from the charging system.

An **ammeter** shows whether the battery is charging or discharging. When the car is running, the gauge should show a slight edge to the charging side. When the motor is off but electrical components are running (heater, lights, etc.) the gauge should show that the battery is being discharged.

An **alternator light** tells you if the charging system is working within the correct range. If the car isn't running and the key is on, the light should be on. If it is on while the engine is running, then the charging system is not working.

INSPECTING THE CHARGING SYSTEM

Before checking the voltage, make sure that ...

- The engine belt is connected and tight.
- The battery terminals are clean.
- Wires going to the battery and alternator are not frayed.
- The battery does not show signs of damage. In addition to a visual inspection, listen to the car as it runs. If the alternator makes a grinding sound, the bearings may be wearing out.

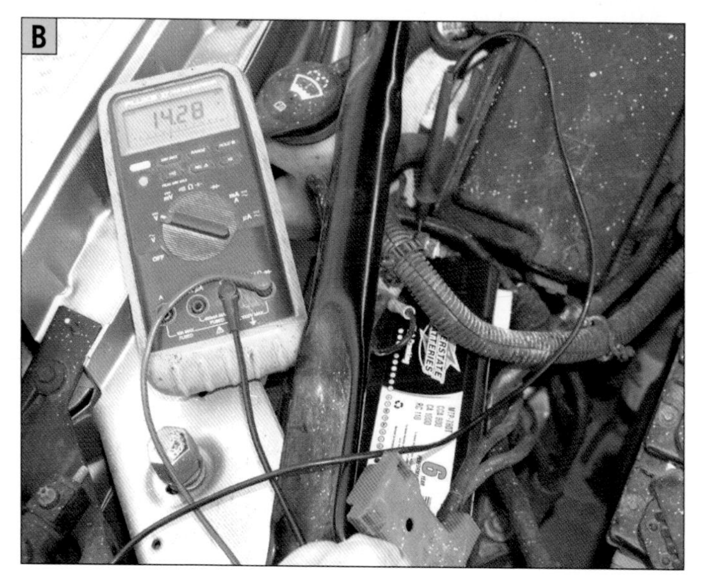

CHECKING THE BATTERY VOLTAGE

1. CHECK THE VOLTAGE WITH THE ENGINE OFF.
   Place your voltmeter across the battery terminals and measure the voltage. It should be around 12 volts (FIGURE A).

2. TURN ON ELECTRICAL COMPONENTS AND CHECK.
   Turn the key on, but don't start the motor. Turn on some of the electrical components (heater, lights, etc.) and recheck the battery. If it has dropped below 10 volts, it may be getting weak.

3. TURN ON THE ENGINE AND CHECK.
   Turn on the engine and let it run. With the accessories off, test the battery again. Your meter should show between 13.5 volts and 14.5 volts (FIGURE B).

4. TURN ON ELECTRICAL COMPONENTS AND CHECK AGAIN.
   Turn on your electrical accessories again and measure the battery voltage again. You should see very little change in the level of charge.

WHAT THIS TEST TELLS YOU

If the battery isn't reading a full charge or drops below 10 volts under load *and* the charging system is showing 13 to 15 volts at the battery while running, the battery is probably wearing out.

If the car is showing too little or too much voltage while the car is running, you may not need a new battery. Check the connections to the battery for corrosion and clean, if needed. Too little output from the charging system usually means a bad alternator, too much charge usually means a bad voltage regulator, and no charge at all indicates a break in the system (belt, cables, or alternator).

How to: Clean the Battery Terminals

Corrosion on the battery terminals and cables is destructive and prevents the battery from functioning properly. Keeping the battery terminals clean will help to keep your car running and performing the way it should. Do a visual check of the terminals on a regular basis, and clean them whenever they show signs of corrosion.

WHAT YOU NEED

- Tool to remove the battery cables from the battery
- Scraping tool or chemical battery cleaner
- Gloves and safety glasses
- Paper towels or rags

1. INSPECT THE BATTERY.
   Look for whitish, chalky buildup around the terminals. This battery has quite a bit of corrosion and needs to be cleaned (FIGURE A). Also check the battery cables for signs of wear.

2. UNHOOK THE CABLES.
   Always disconnect the battery cables before working on or cleaning the battery. Unhook the ground cable first. It is usually black and identified with a minus sign (-). Pull the cables away so they don't accidentally touch the battery posts (FIGURE B).

CLEANING PRODUCTS

Corrosion on battery terminals and cables can be removed manually or with chemical cleaners. You can purchase an inexpensive tool designed to scrape the battery posts clean, or you can use steel wool or emery paper to clean the contacts. Commercially available battery cleaning sprays can also be used.

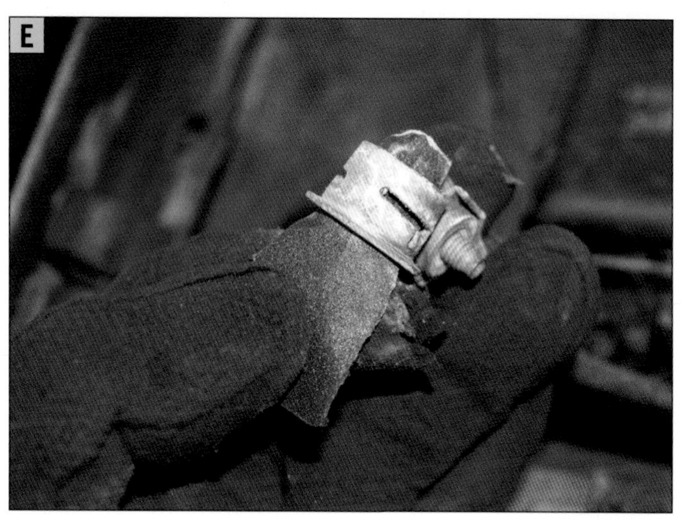

3. REMOVE CORROSION FROM TERMINALS AND POSTS.
   Clean the inside of the battery terminal and the outside of the battery posts with your tool or other cleaner.

   This battery cleaning tool fits over the top of the post and scrapes off corrosion (FIGURE C).

   If you use a chemical cleaner, follow the manufacturer's instructions and wear safety protection. Wipe the terminals and battery thoroughly after you are done (FIGURE D).

4. REMOVE CORROSION FROM CABLES.
   You can use a piece of emery paper to clean the inside of the cable (FIGURE E). Make sure you have a nice, fresh surface when you are done.

5. REINSTALL THE CABLES.
   Wipe the terminals clean with paper towel or rags and reinstall the cables. Your cables need to be tight so they don't move, but not so tight that they break or twist the terminals.

TIP

A little petroleum jelly on the terminals keeps the air out of the connection and helps to prevent further corrosion.

How to: Change a Battery

Before changing your battery, check the charging system and battery connections to make sure your problem isn't due to a faulty connection.

WHAT YOU NEED

- Replacement battery
- Tools to remove battery hold-downs and brackets
- Supplies for cleaning the terminals
- Gloves and safety glasses
- Paper towels or rags
- Petroleum jelly (optional)
- Side-post battery wrench (optional)

A

B

C

1. LOCATE THE BATTERY.
   Batteries are often located in the front of the engine compartment, like this one (FIGURE A). Check your owner's manual for the location of your battery.

2. UNHOOK THE GROUND CABLE AND THEN THE POSITIVE CABLE.
   Always unhook the ground cable first. It is usually black and identified with a minus sign (-). This battery is a side-terminal style battery, which connects the cables to the battery with a bolt (FIGURE B). After unhooking the ground, unhook the positive cable. It is usually red and marked with a plus sign (+).

3. REMOVE BRACKETS AND OTHER OBSTRUCTIONS.
   Remove any brackets that may keep you from accessing the battery. This car has a structural bracket that must be removed to get to the battery (FIGURE C). We also had to unhook the fuse box mounted behind the battery and pull it out of the way.

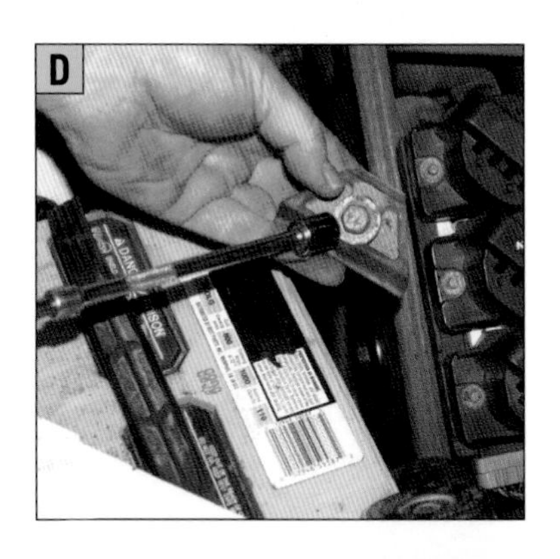

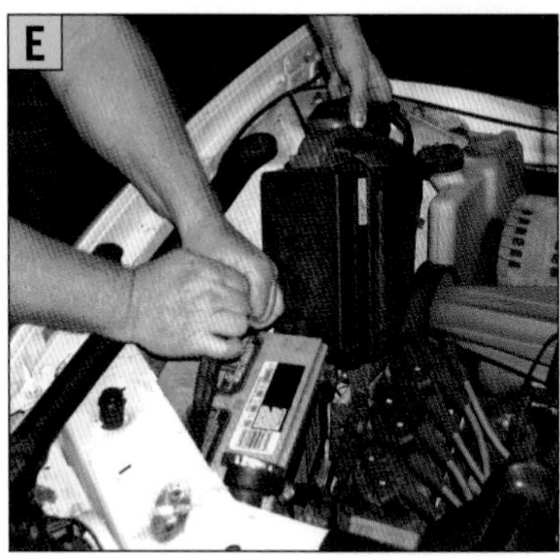

4. REMOVE THE BATTERY HOLD-DOWN BRACKET.
Most batteries have a hold-down bracket to keep them in place. Some go across the top of the battery, while others, like this one, hold down a small ledge at the bottom of the battery (FIGURE D). Locate the hold-down bracket and remove it. Do not forget to reinstall it after changing the battery.

5. REMOVE THE BATTERY.
The battery is very heavy and you may need assistance to remove it. In this case, two people were needed—one to hold the fuse box out of the way and one to pull the battery (FIGURE E). If you don't feel you can lift the battery, find help.

6. CLEAN THE BATTERY COMPARTMENT.
With the battery out, wipe the dirt out of the battery tray area with paper towels or rags. If you see evidence of battery acid or any other contaminant, wear gloves to prevent contact. If needed, gently rinse the area with clean water.

7. PREPARE THE NEW BATTERY.
Prep the new battery by scratching the oxidation from your battery cables and new battery contacts. Apply a thin layer of petroleum jelly to the terminals and bolts to prevent oxidation on the terminals (FIGURE F).

8. INSTALL THE NEW BATTERY.
Installing the new battery is the reverse of removing the old one. Check that you haven't set the new battery on any wires or cables, and secure it tightly. Remember to replace the hold-down bracket. Install the positive cable first, followed by the ground.

9. RESET YOUR CAR.
With your battery in place, reset your clock, preset radio stations, and any other electronic devices that utilize memory.

10. RECYCLE YOUR OLD BATTERY.
Old batteries contain toxic elements and must be disposed of properly. Most battery retailers will also accept batteries for recycling, as do local recycling centers. You may be charged a fee to cover recycling costs.

CHAPTER 12

THE BRAKING SYSTEM

IN THIS CHAPTER

How the Braking System Works

Braking systems haven't changed too dramatically in the history of automobiles. Like the first cars, most systems today use friction to slow the car down. Simple levers have been replaced with hydraulics, which apply greater pressure to the brakes for better stopping power, but the basic principle remains the same.

Conventional brakes work by applying a brake pad to a braking surface, which generates friction. The car slows as its forward motion (kinetic energy) is converted into heat energy. Because the friction between the two surfaces creates heat, the braking system is designed to wear slowly and dissipate heat quickly. The brake system needs regular maintenance to keep you safe on the road.

Most cars have three braking systems: the front or primary brake system (most of your braking comes from the front brakes), the rear or secondary brake system, and an emergency brake system, which is usually mounted with the secondary system.

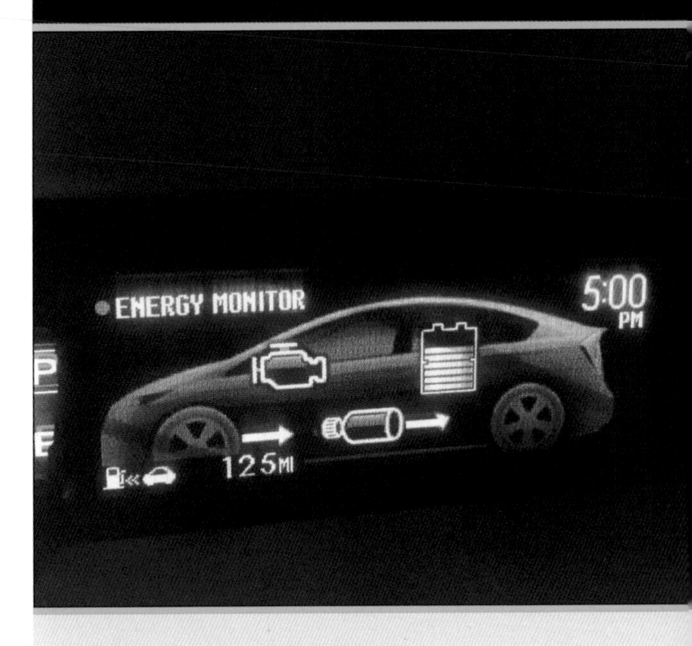

REGENERATIVE BRAKING

Electric and hybrid cars are usually equipped with conventional brakes, but they can also take advantage of *regenerative braking*. Instead of converting kinetic energy to heat energy through friction, regenerative brakes run the power in reverse through the electric motor, which acts as a resistance to the motor, slowing the car. Cars that use regenerative braking rely less on the conventional hydraulic systems, which makes them last longer and run cooler.

TYPES OF BRAKES

Disc Brakes A disc brake uses hydraulic power to push the brake pads against a rotor that rotates in the same direction as the wheel. Disc brakes use higher pressures and more fluid than drum brakes. They don't have springs to retract them away from the rotor, so they drag slightly against the rotor until needed. Most cars have this type of braking system.

Drum Brakes Drum brakes use friction pads to push against the sides of a drum. Like disc brakes, they are hydraulic, but they use less pressure and fluid to operate. Unlike disc brakes, drum brakes retain some hydraulic pressure, so they use a series of springs to retract them from the drums to prevent drag. This makes them more complex to refurbish.

Mechanical or Emergency Brakes The emergency brake system is a mechanical backup in case the hydraulic system fails. By pulling a lever or pressing the foot switch, you can engage the brakes manually. The mechanical system usually has a method of holding the brakes in an "on" position for emergency situations or if you have to park on an incline and you need to set the brakes to hold the car in place.

BRAKE COMPONENTS

Brake Pads The brake pads, also called brake shoes, are coated with a friction material that is designed to wear away by changing the motion of your car into heat, and this stops the car. They can be made of many different types of materials including natural and manmade fibers, fibers mixed with metals (semi-metallic), ceramics, and other components. The type of brake pad you need is determined by the car manufacturer. Only change the type of pad you use if you have changed the use of your car. (For example, high-performance vehicles usually have upgraded brakes and brake pads.) The brake pads are designed to wear away, and they need to be replaced at regular intervals.

Rotor The rotor is the big, heavy piece that the brake pads push against to stop the car. They are usually made of metal and may be coated with ceramic or other material to help them resist or dissipate heat faster. Most primary brake rotors are vented, which means they have holes running through the outside of the rotor. These help the rotor dissipate heat by allowing airflow between the metal braking surfaces. Some rotors have holes through them and slots cut in the braking area (called "cross-drilled" or "slotted" rotors). These help the rotor remove heat faster, remove gas, and remove water picked up off the road from the braking components. They are usually used on performance cars.

Master Cylinder The master cylinder has a reservoir that holds the fluid and regulates the flow of the hydraulic brake fluid. When you push the brake pedal, the master cylinder pushes the brake fluid out to the primary and secondary brake systems. It keeps the two systems separate, so if one system fails, you still have the other.

Calipers and Brake Cylinders These components are at the end of the hydraulic system. They apply the force to push the brake pads against the braking surface. Calipers use one or more pistons to press a pad against the brake rotor, which pushes the caliper outward and squeezes the two pads against the rotor. A brake cylinder pushes both brake pads out at the same time, and uses springs to retract them back into position.

Power Brake Boosters A power brake booster uses vacuum, hydraulic, or electric power to increase the amount of pedal pressure you can apply to the brakes. It is usually mounted between the brake pedal and the master cylinder. Your foot applies around one-tenth of the pressure power brakes can apply, which allows manufacturers to design power brakes to work better with increased ease for the driver.

Anti-Lock Brakes (ABS) Most cars have anti-lock brakes to help you slow safely in slippery conditions. When your brakes lock up, you are no longer using the brake pads to convert forward motion into heat; you are using the surface of your tires. ABS can sense if a wheel has stopped prematurely and will release brake pressure to that wheel, enabling it to move and transfer energy back to the brake pads. The ABS is mounted between the master cylinder and the four braking points. If the ABS engages, you will feel the brakes pulse as the brakes are applied and then released by the ABS. This pulsing is normal, and by pulsing the brakes the ABS can slow the car much faster than skidding.

Hydraulic and Mechanical Lines Hydraulic brakes use a series of flexible and non-flexible lines to carry the high hydraulic pressures from the master cylinder to the calipers or brake cylinders. Emergency mechanical brakes generally use cables attached to the secondary brake system to engage.

Common Braking System Problems

It's important to recognize the signs of brake problems and to know what to do if they occur. The ability to stop is more important than the ability to go. This list is not all inclusive, but it covers some of the more common brake problems.

NOISY BRAKES

The brakes should not make any noise when operating. If they squeal or make a metal-on-metal grinding sound, the brake pads are probably worn and need to be replaced. It is important to replace worn pads quickly, as they can grind or imbed friction material on the rotors. Brakes may also squeal if they become glazed. Brake glaze occurs when heat hardens the metal rotor or pads, making them shiny and smooth, a condition that can reduce their effectiveness and require replacement.

VIBRATION

You may notice the car shuddering as you apply the brakes and the shuddering slowing as the car slows down. This vibration is usually caused by a crack in the rotors, uneven wearing of the pads or shoes, or the transfer of the friction material onto the rotor surface.

If you find a crack in either the metal or ceramic coating of the rotor or drum, the rotor or drum will need to be changed. Replace unevenly worn brake pads, and have the rotor turned (resurfaced) to even out the surface. Embedded brake pad material can be machined off the rotors if they are not already too thin.

It is possible, although less likely, that a warped rotor is causing the vibrations. Rotors can become warped if too much material is removed when the rotor is turned, and a warped rotor needs to be replaced. Rotors generally should be turned when a regular brake job is performed to clean and even out the braking surface.

PULLING TO ONE SIDE

If you notice the car pulling to one side when you apply the brakes, it is likely due to contamination in the brake components, or a faulty caliper or brake cylinder. It can also happen if you have differently sized tires, such as when you're driving on a temporary spare. Worn suspension parts can also cause this problem.

SOFT OR SPONGY BRAKES

You may begin to notice over time that you have to push harder on the brake pedal to engage the brakes and bring the car to a stop. If the brakes are still working, the usual suspects include low fluid level, worn pads, worn out or dirty fluid, or a failed booster.

As disc brakes wear, they don't retract like drum brakes, so the level of fluid decreases. If you go too long without adding more fluid or changing the fluid, they can give you excessive brake pedal travel.

If the brake pedal goes to the floor but you have little or no braking action, the brakes have lost the ability to produce pressure. This can be caused by a break or leak in the lines or brakes, or a failure of the master cylinder or distribution block.

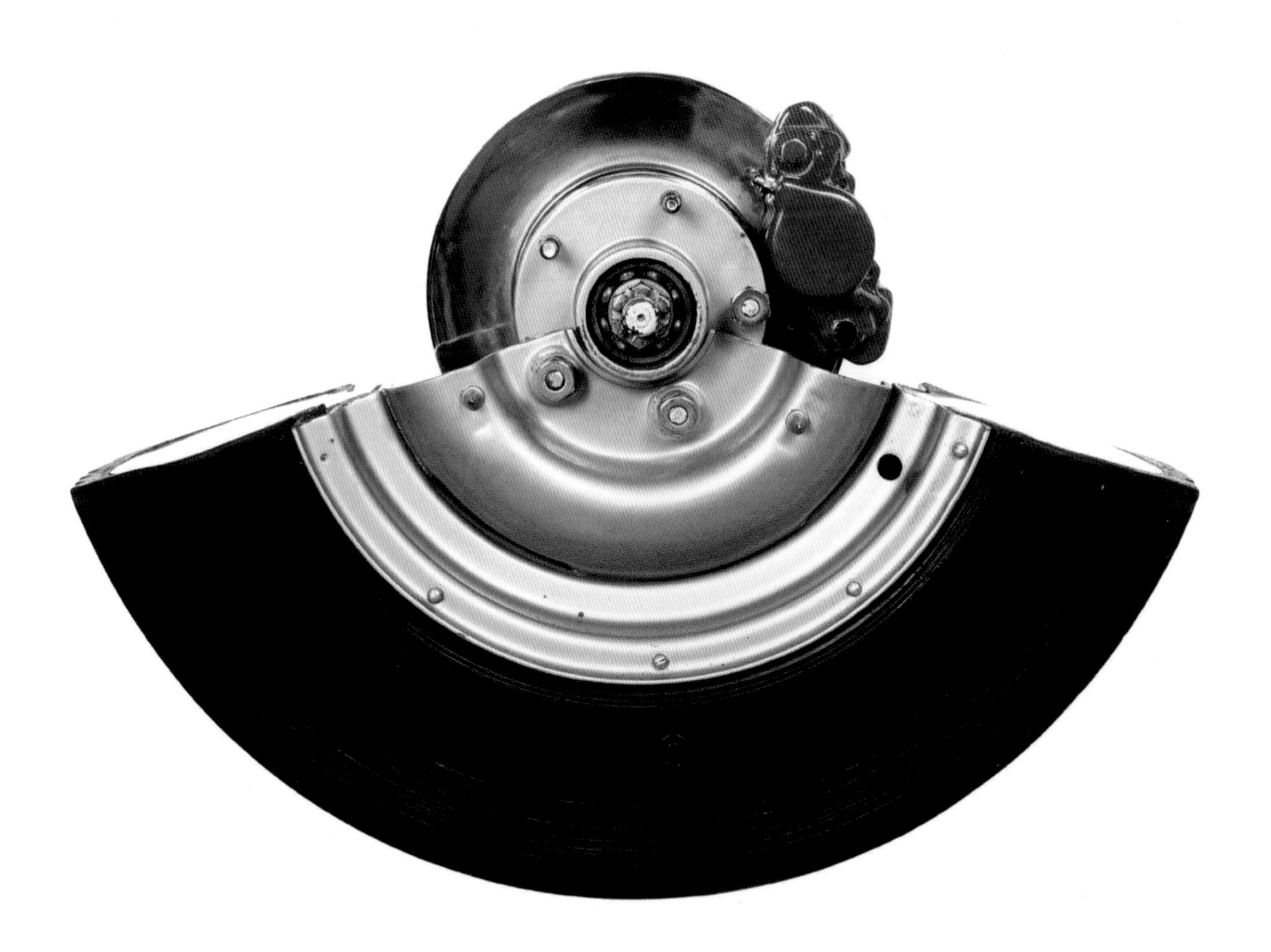

How to: Inspect the Brakes

You should check your brakes or have them inspected about every 10,000 miles or 15,000 kilometers. You can inspect disc brakes on your own, but you may want to leave the more complicated drum brakes to a professional. When inspecting the brakes, be cautious with heat, brake dust, and brake fluid. Wearing gloves and eye protection is strongly recommended.

WHAT YOU NEED

- Jack and jack stands
- Safety gloves and eye protection
- Torque wrench and socket

1. RAISE THE VEHICLE AND REMOVE A WHEEL.
   Check the brakes one wheel at a time, and when the brakes are cool. Removing the wheel will give you the best access for inspecting the brakes.

2. INSPECT THE ROTOR.
   The rotor should be smooth, but not glossy. You may notice some very tiny ridges in the rotor, or a ridge on the outside edge. These are normal.

   Look for discoloration. This may indicate overheating, embedded brake pad material, or a rotor that is made of inferior materials.

   Look for signs of uneven wear. If you can, spin the rotor by hand and listen to the sound. A disc brake does not retract, so you should hear an even sound of the pad lightly touching the rotor. If it sounds uneven, you may have a problem with the brakes.

   Look for cracks or gouges in the rotor. If you find any, the rotor will need to be replaced.

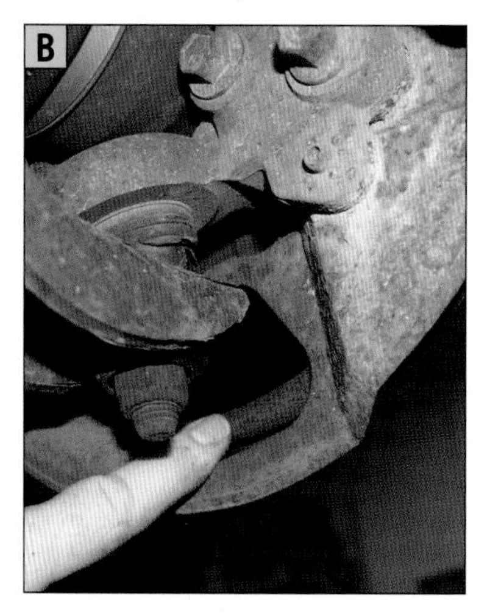

Viewing the back side of the rotor can be difficult due to the brake dust shield. This car has a small opening for inspecting the back side (FIGURE B).

This rotor has friction material from the brake pads lodged in it. This causes the brakes to slow the car unevenly. The rotor will need to be turned to remove the friction material (FIGURE C).

This rotor is heavily grooved due to the brake pad being worn down to the metal base. It needs to be replaced (FIGURE D).

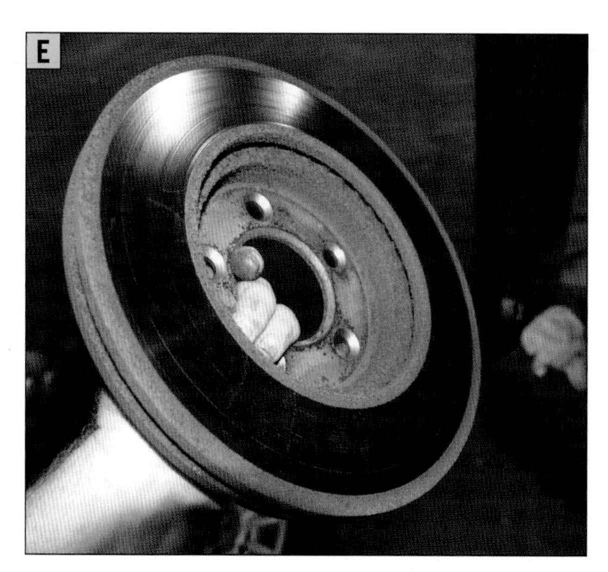

This rotor is glazed. Glazing hardens the surface metal, making the pad and rotor less effective, and causing squealing brakes. A glazed rotor can be resurfaced if the rotor is thick enough (FIGURE E).

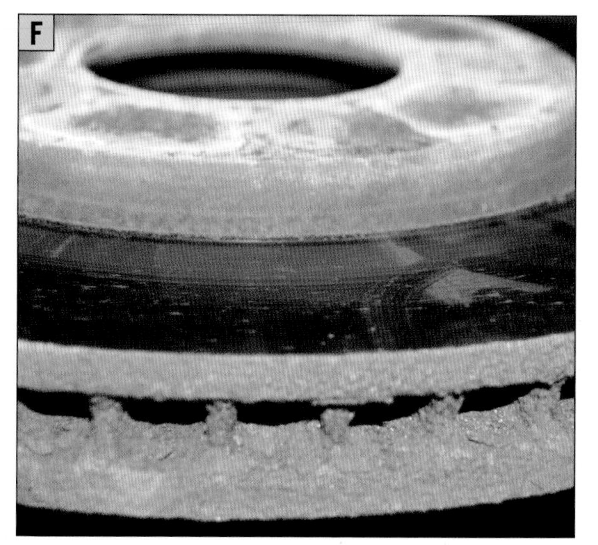

A rotor with a very deep ridge may need to be replaced. When a rotor is turned, it typically takes about a postcard-thick amount of metal off each side of the rotor. If the ridges are deeper than this you will want to have it checked (FIGURE F).

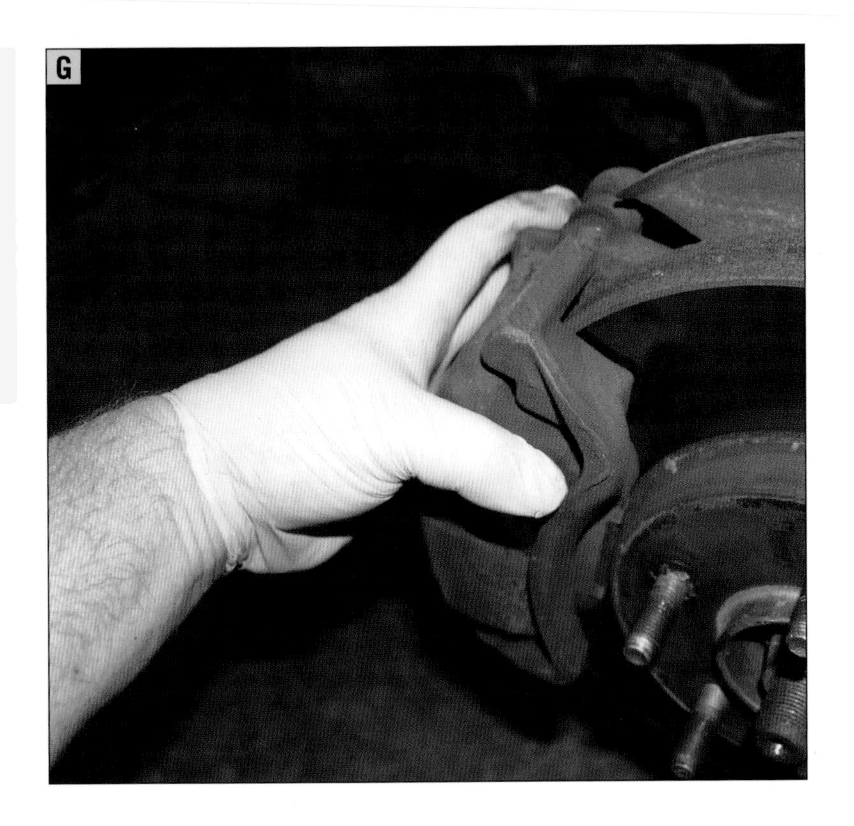

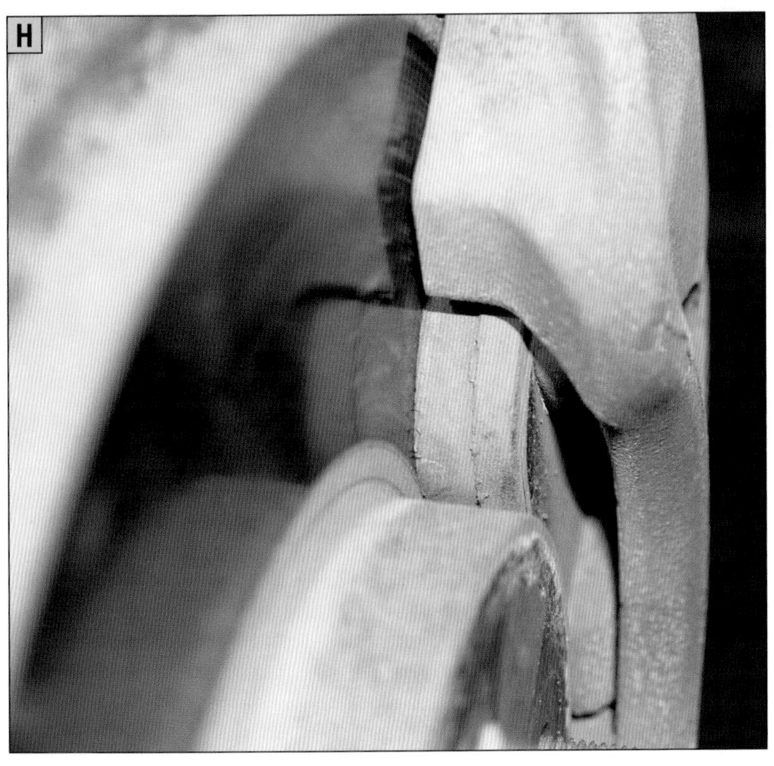

3. **INSPECT THE CALIPER.**
   If the caliper is cold, wiggle it back and forth to make sure it isn't loose and the mounts have not worn out. Look for signs of leaking brake fluid inside and outside (FIGURE G).

4. **INSPECT THE BRAKE PADS.**
   There should be a minimum of 1/8 inch to 3/16 inch of material on the brake pads between the rotor and the metal backing plate. A good rule of thumb is that the pad material should be thicker than the backing plate, as shown.

   If one pad is wearing more than the other, there may be a problem with the piston or slide pins on the caliper. Look for even, flat wear (FIGURE H).

5. **INSPECT THE BRAKE LINES.** Check the fluid lines coming in to the brake caliper for abrasions, rust, or kinks (FIGURE I).

6. **REINSTALL THE WHEEL AND TORQUE THE LUG NUTS.** Make sure you properly reinstall the wheel before removing the safety stands and moving to the next wheel.

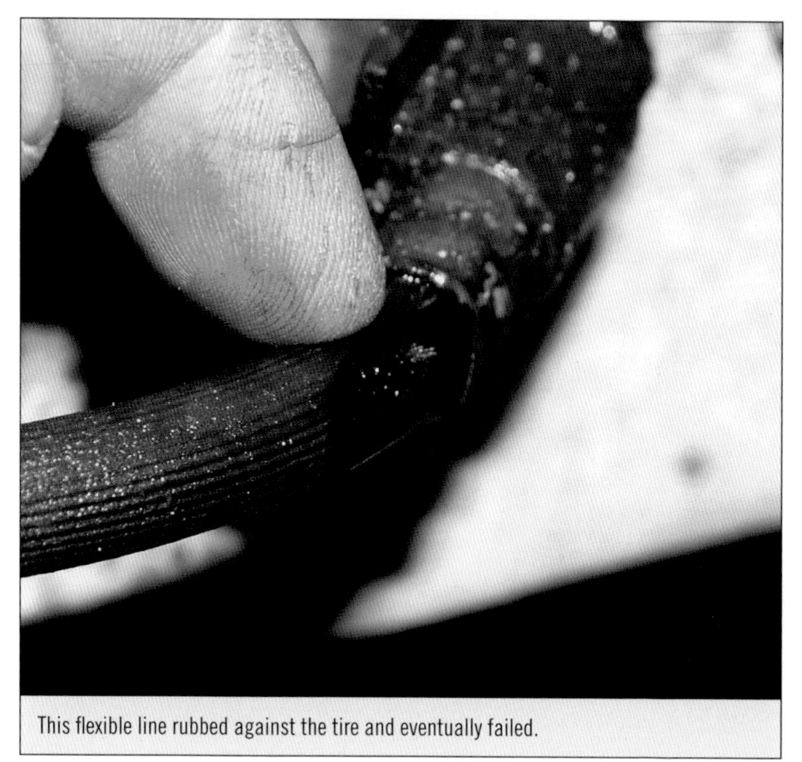

This flexible line rubbed against the tire and eventually failed.

How to: Change Brake Pads and Rotors

Changing the pads and rotors on most front-wheel-drive cars is not terribly difficult. However, it's important to keep in mind that all cars are designed differently. Have a shop manual for your vehicle on hand, and only change the pads and rotors on one side of the car at a time so you can use the other side as a reference.

WHAT YOU NEED

- Jack and jack stands
- Safety gloves and eye protection
- New pads and rotors
- Torque wrench and socket
- Tools to remove the caliper and caliper mount
- Disc brake spreader or a C-clamp and block of wood
- Brake assembly grease
- Turkey baster (if needed)

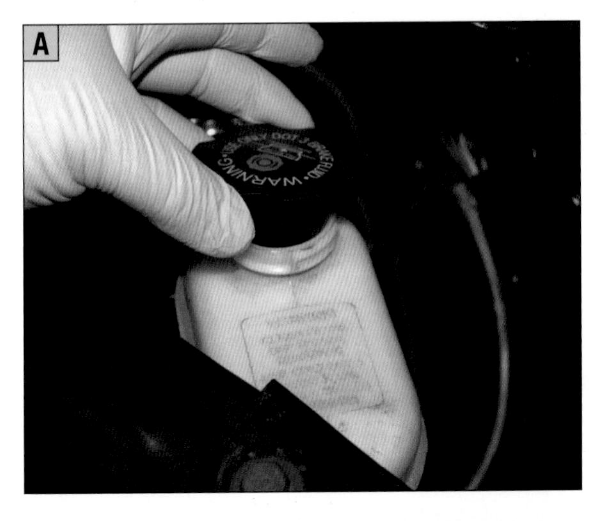

B

This car has glazed front brakes.

1. RAISE THE VEHICLE.
   Raise your vehicle safely using the jack and car stands.

2. REMOVE ONE WHEEL.
   Use the torque wrench and socket to remove the lug nuts and pull one wheel from the vehicle.

3. OPEN THE MASTER CYLINDER.
   Twist off the master cylinder cap to allow air to escape when you compress the caliper piston (FIGURE A).

4. INSPECT THE ROTOR.
   Look for deep grooves, discoloration, and glazing (FIGURE B).

TIP

Most rear-wheel-drive cars have drum brakes or disc brakes that ride on a spindle. These designs are complicated and require special equipment. You may want to leave changing the pads and rotors on these types of brakes to a professional.

C

D

E

5. REMOVE THE CALIPER.

There are two types of brake calipers, floating and fixed. Most brakes are floating, like this one, which means they move on a set of slides and have one or more pistons on one side to push the caliper against the rotor. A fixed caliper has one or more pistons on both sides, and it is mounted so it can't move. The pistons push the pads from both sides rather than from one side (FIGURE C).

If you are doing rear brakes, you may need to remove the emergency brake cable to get the caliper out of the way to remove the rotor.

6. SECURE THE CALIPER AND REMOVE THE BRAKE PADS.

There are usually two bolts holding a floating caliper. Remove these two bolts and secure the caliper up and out of the way. Be sure not to put any strain on the flexible brake line (FIGURE D).

The brake pads may or may not come out by hand. A fixed pad may need to be removed with a tool.

7. INSPECT THE CALIPER.

With the old pads out, you can inspect the caliper piston. Check the seal between the caliper and piston for breaks, tears, and leaks. Clean the loose rust and debris out of the caliper area (FIGURE E).

8. **REMOVE THE CALIPER MOUNT.**

   It may be necessary to remove the caliper mount or other brackets to get the rotor off of the car. These may be hard to remove as they typically are installed with a lot of torque and sometimes manufacturers use thread locker to hold them in place (FIGURE F).

9. **REMOVE THE ROTOR.**

   Some manufacturers use little clips to hold the rotor to the car while it is being assembled. Save these if you can. They can be reused to hold the rotor in place when reassembling the brakes (FIGURE G).

   The rotor may stick to the metal hub on which it's mounted. Use a mallet and gently knock the rotor loose and remove it from the car.

10. **INSTALL THE NEW ROTOR AND BRACKETS.**

    Install the new rotor, using your clips if you still have them. Reinstall the brackets and check your shop manual for the correct torque specification for the brackets (FIGURE H).

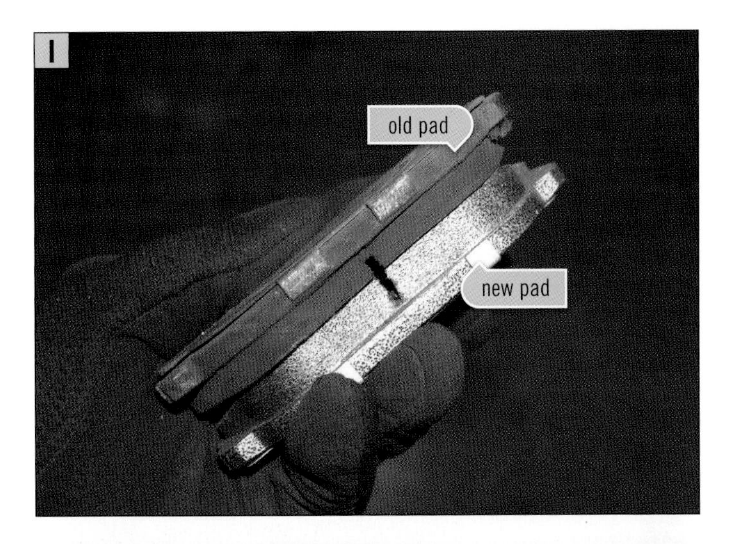

11. INSPECT THE OLD PADS.

Check the old pads for uneven wear, both on the individual pad and against the other pad. Here is a side-by-side comparison of old and new pads. The old pads did not have much wear, but were replaced due to noise (FIGURE I).

12. PUSH THE PISTON BACK INTO THE CALIPER.

Disc brakes don't retract like drum brakes, so to install the new thicker pads, the piston has to be pressed back into the caliper. You can purchase a special tool to do this, or you can use a big C-clamp and a piece of wood (FIGURE J). The wood helps to spread the pressure around the piston and keep it going straight.

Take your time and press the pistons back slowly. You're pushing the brake fluid back into the master cylinder and you don't want to rush. Once the piston is flush with the seal, you are done.

13. GREASE THE SLIDE PINS.

The slide pins are used to move a floating caliper in and out. Your caliper will mount to these pins. Remove them, clean off the old grease, and apply new brake grease. Inspect the pin covers for any tears and replace if necessary (FIGURE K).

CAUTION

One of the big reasons brakes fail is due to improper seating. Different brakes require different methods to seat them and make sure they don't embed material into the new rotors. Follow your brake manufacturer's recommendation, which can usually be found on the box with the pads.

14. GREASE AND INSTALL THE NEW PADS.

To reduce the amount of noise coming from your brakes, apply brake assembly grease to the brake pad. Grease the back of the pad that touches the piston, the metal tabs where the back of the brake pad touches the caliper, and the pad retainers (FIGURE L).

Install the new pads in the caliper. If they are installed with clips, use new clips provided by the pad manufacturer.

15. REINSTALL THE CALIPER.

Reinstall the caliper and slide pins and tighten to your manufacturer's specifications. Spin the wheel and make sure everything turns freely. The pads may not touch the rotor yet since you retracted the piston (FIGURE M).

16. CHECK THE FLUID LEVEL AND REINSTALL THE WHEEL.

Since you pushed fluid back up into the master cylinder, the fluid level may be high. If it is, remove the excess with a turkey baster.

Once the level is good, reinstall the wheel and torque the lug nuts.

17. LOWER THE CAR AND REPEAT WITH REMAINING BRAKES.

After lowering the car, begin the process again with the brakes on the opposite side of the car.

18. SEAT THE BRAKES.

Once both or all the brakes have been replaced and your master cylinder level has been set and closed, pump the brakes several times to reset the brake pads against the rotor. Do this before test driving the car.

TURNING YOUR ROTORS

Whenever you change your brake pads, you should have your rotors turned or install new rotors. This is important because the brake pads and rotor are designed to press flat against each other, and when brake pads wear, they don't wear flat. If you change your pads but don't replace or turn the rotors, you will have a flat pad pressing against an angled rotor. Until the pads wear to match the worn rotor, your brakes will not perform optimally.

If you don't want to replace your rotors, you may be able to have them turned by a mechanic. Turning (also called *resurfacing* or *machining*) is a process in which a very small amount of material is removed from the rotor surface to make it even again. This can only be done if your rotors are not too worn. If your rotors are too thin, they will need to be replaced.

You can measure the thickness of the rotor, but doing so properly requires special tools. If you are changing pads, it's best to let a mechanic measure the rotors. They'll have the right tools and the know-how, and you can still save some money doing the rest yourself.

How to: Check and Fill Brake Fluid

Hydraulic brakes have a reservoir of brake fluid mounted on top of the master cylinder. As you press on the brakes, fluid is drawn from the reservoir, and when you let off the brakes, the fluid is returned. If the fluid level gets too low, the brakes could draw air into the brake system, which could prevent the brakes from operating properly. The brake fluid can also become contaminated with soot from use. Brake fluid should be checked and changed regularly.

WHAT YOU NEED

- Brake fluid
- Latex or nitrile gloves
- Safety glasses
- Turkey baster (optional)

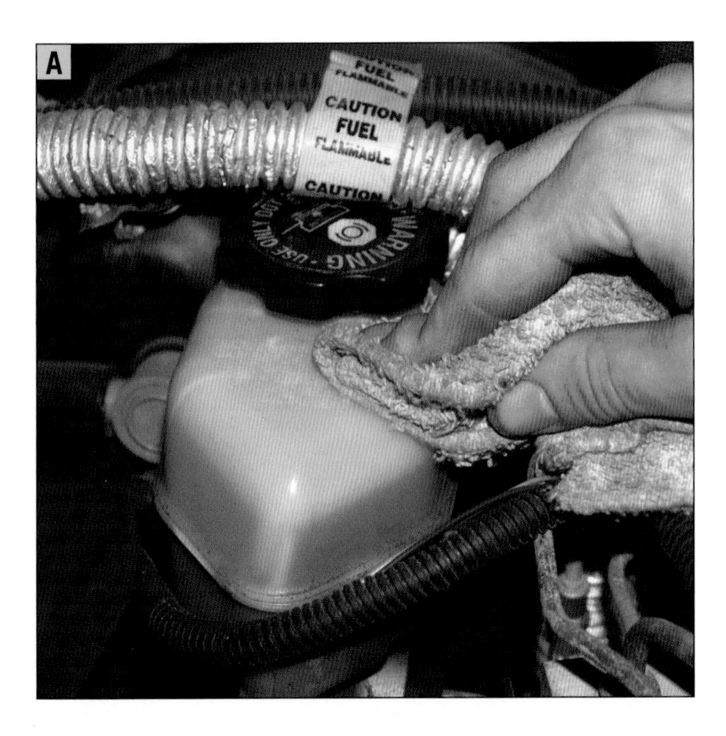

CAUTION

Brake fluid is corrosive and should not come in contact with your skin. Wear latex or nitrile gloves and safety glasses when handling brake fluid.

1. LOCATE THE MASTER CYLINDER RESERVOIR.
   The master cylinder is usually mounted near the brake pedal in the engine compartment. Look for the minimum and maximum level lines on the side of the reservoir. Under normal driving conditions, the fluid level should stay near the maximum line and not go down too much. If the level is down, there may be a leak in the brake system, or the disc brake caliper pads may have worn down and need to be replaced.

2. WIPE OFF THE MASTER CYLINDER.
   Before you add fluid or open the reservoir to check it, it's a good idea to wipe the top and cap clean so you don't get any dirt down in the master cylinder (FIGURE A). Outside contaminants can clog the hydraulic system.

BRAKE FLUID DESIGNATIONS

All brake fluids carry a standard designation from the Department of Transportation. The most widely used designations are DOT 3, DOT 4, and DOT 5. DOT 3 and DOT 4 can be mixed together, but DOT 5 is made of different materials and you cannot mix it with other fluids. Check the cap of the master cylinder or your owner's manual for the kind of fluid to use in your system.

The fluid in this car has a lot of soot buildup, so it needs to be changed.

3. INSPECT THE FLUID.
   Look down in the reservoir and inspect the fluid or use a turkey baster to draw out some fluid, if needed (FIGURE B). Brake fluid should be clear or amber in color. If your brake fluid is dirty, bleed the brake system before adding more fluid.

4. ADD FLUID, IF NEEDED.
   If your brake fluid is clean but low, add fluid to the maximum level mark on the side of the reservoir and replace the cap (FIGURE C). If you put in too much, use a turkey baster to remove the fluid until it is at the correct level. Low fluid levels can indicate a leak in the system, so inspect the brakes for wear or leaking.

TIP

You can top off dirty fluid temporarily, but you should bleed the system and change the fluid as soon as possible to get the contaminants out of the system.

How to: Bleed the Brake System

You need to bleed the brakes if there is air in the brake lines or if the fluid needs to be changed. You can use one of three different methods: the gravity method, pressure bleeding, or vacuum bleeding. The gravity method is the simplest, and relies on gravity to drain the fluid and air. Pressure bleeding uses the brake pedal to force out the air and fluid, and vacuum bleeding uses a vacuum pump on the end of the line to draw out the fluid and air.

WHAT YOU NEED

- Brake fluid
- Line wrench
- Clear plastic tubing
- Clear jar (for gravity and pressure bleeding)
- Vacuum pump (for vacuum bleeding)
- Grease (optional)
- Rust-penetrating spray (optional)
- Jack and jack stands (if needed)
- Latex or nitrile gloves and safety glasses

GRAVITY BLEEDING

Gravity bleeding is the easiest way to bleed the brakes. You simply open the bleeder screws and allow gravity to do the rest of the work. Be sure to recycle the used brake fluid properly when done.

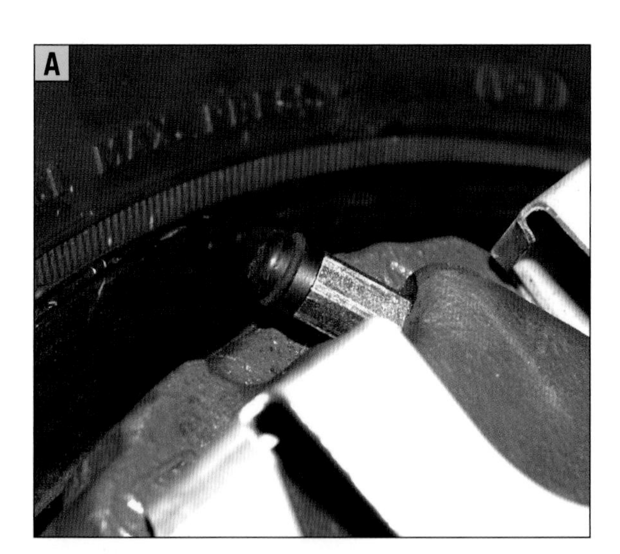
A

1. RAISE THE CAR AND SECURE IT.
   If you need room to work under your car, carefully raise the car and secure it with jack stands.

2. LOCATE THE BRAKE BLEED SCREWS.
   The bleed screws are usually located on the top of the calipers and brake cylinders. They look like a thick hex screw with a hole in the end. Sometimes they have a rubber cap on the end to keep out dust (FIGURE A). If needed, apply a little penetrating spray on the bleed screws to help loosen them.

CAUTION

Brake fluid is corrosive and should not come into contact with your skin. Wear latex or nitrile gloves and safety glasses when working with brake fluid.

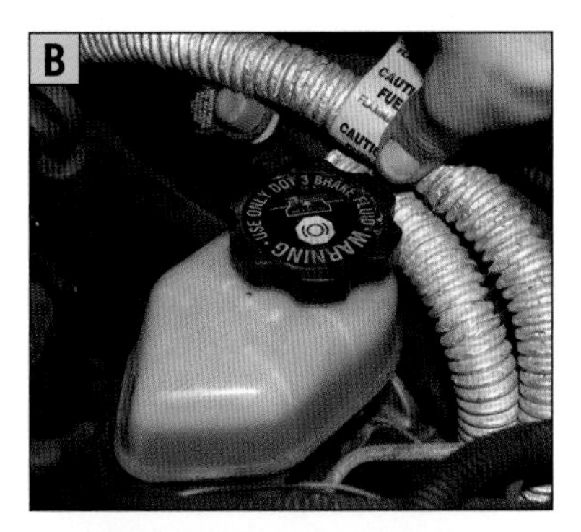

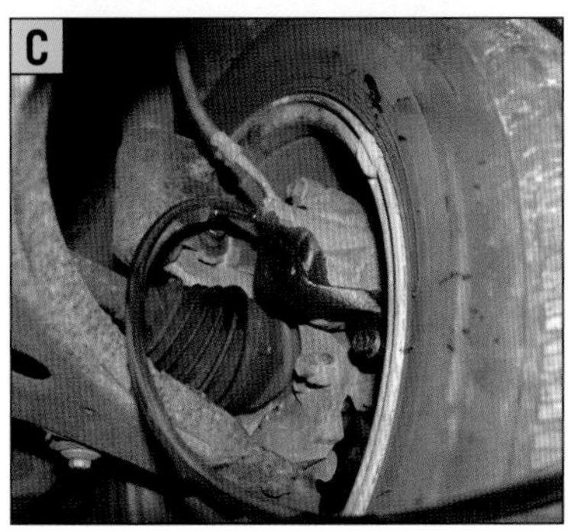

3. **PREPARE THE MASTER CYLINDER.**
   Top off the master cylinder. For gravity bleeding and vacuum bleeding, you need to remove the cap. For pressure bleeding, you need to close the cap. Regardless of which method you use, it is important that you do not let the master cylinder go empty. If it gets too low, air can be drawn into the brake system (FIGURE B).

4. **CONNECT THE TUBING AND OPEN THE BLEED SCREWS.**
   Place one end of the clear tubing in a catch jar, connect the other end to each bleed screw to be bled, and carefully open the bleed screws. The tubing will spin with the screw. It doesn't take much (quarter to a half turn usually) to open the screw and start letting the brake fluid out. Let the system continue to flow until you see the fresh new fluid coming out (FIGURE C).

   If doing only the front or rear brakes, open both screws on that system. If doing both front and back, open all four.

   Gravity feeding can take a while. Be sure you keep an eye on the fluid level in the master cylinder, and top off the fluid as it draws down.

5. **CLOSE THE BLEED SCREWS AND TOP OFF THE MASTER CYLINDER.**
   When you begin to see clean, fresh fluid coming through the hose, you can close the bleed screws and remove the hose. Top off the master cylinder and wipe up any spilled fluid.

6. **CHECK YOUR BRAKES BEFORE DRIVING.**
   With your vehicle running and stationary, check your brakes for a firm pedal before operating the vehicle. Allow the vehicle to move slowly and check the brakes several times. If you have any concerns, do not operate the vehicle, and take it to a professional as soon as possible.

7. **RECYCLE THE USED BRAKE FLUID.**
   Be sure to dispose of old brake fluid properly by putting it in an appropriate container and taking it to an auto parts dealer or recycling station. Never reuse old brake fluid.

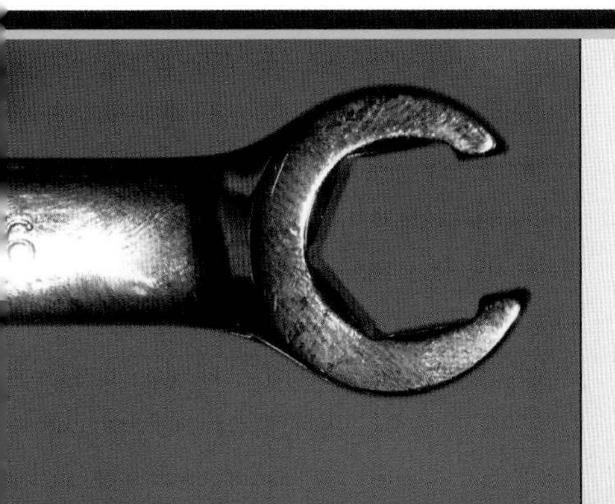

USING A LINE WRENCH

Use a line wrench to loosen bleed screws. It has an opening on the end to get around hoses and brake lines. It is also thicker than a standard wrench and holds on six sides. A regular twelve-sided wrench may strip the soft metal of the bleed screws.

PRESSURE BLEEDING WITH TWO PEOPLE

The pressure bleeding method uses the brake pedal to help force the air and fluid through the system. It's easiest to do this job with the help of another person.

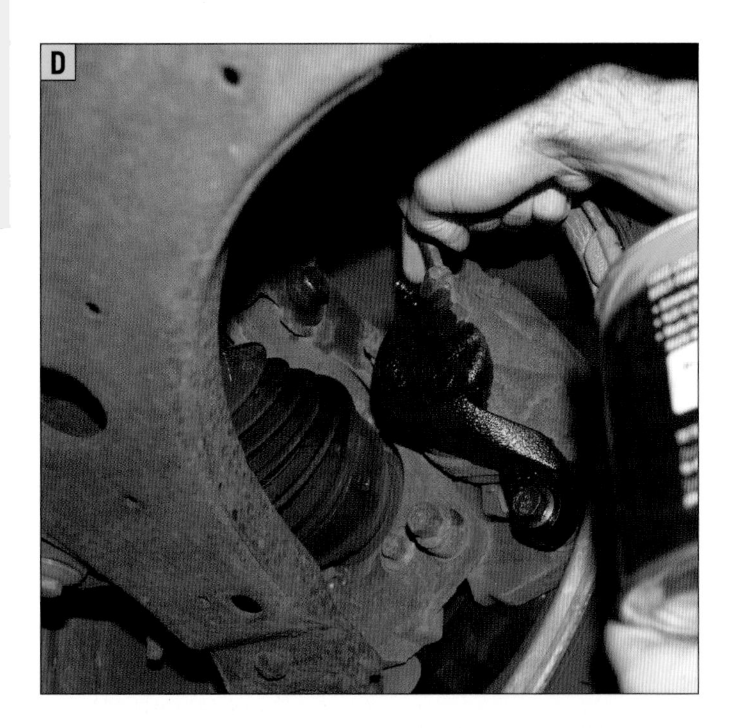
D

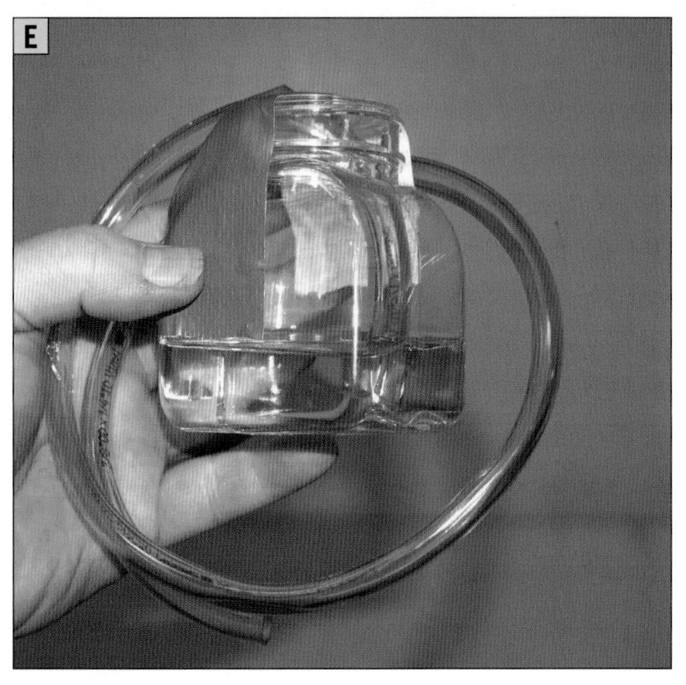
E

1. **PREPARE VEHICLE FOR PRESSURE BLEEDING.**
   Follow the instructions for Gravity Bleeding, steps 1 through 3.

2. **APPLY GREASE TO THE BLEED SCREWS. (OPTIONAL)**
   Applying a little grease around the screw base and the opening at the top helps to seal the threads and tube when you release the brake pedal (FIGURE D).

3. **PREPARE THE CATCH JAR.**
   Partially fill the jar with clean brake fluid and place one end of the tubing in the jar, below the fluid level (FIGURE E). This will allow the bubbles to escape, and if anything gets pulled back up in the tube it will be fresh brake fluid.

4. **ATTACH THE TUBE TO THE BLEED SCREW.**
   Attach the other end of the tube to the first bleed screw, making sure it is snug. Begin with the brake that is farthest from the master cylinder. This is usually on the passenger side. If doing all four brakes, begin with the rear passenger-side brake.

5. PUSH THE BRAKE PEDAL AND OPEN THE BLEED SCREW.

Have one person slowly press the brake pedal, taking about two seconds to reach the floor.

As the pedal is being pushed, have the second person open the bleed screw, allowing the fluid to flow through the tube and into the jar (FIGURE F). Close the screw just before the pedal reaches the floor.

After the bleed screw is closed, the person pushing on the pedal can release the pedal.

6. REPEAT UNTIL FLUID RUNS CLEAN.

Repeat this process of pushing the pedal, opening the screw, and closing the screw until new, clean fluid comes through the tube.

7. REPEAT WITH THE NEXT BRAKE LINE.

Once you have clean fluid coming through the tube, close the bleed screw and repeat the process with the next brake line until all brakes have been bled. Make sure you keep the master cylinder topped off and reinstall the cap after each top-off.

8. CLEAN UP AND TEST THE BRAKES.

Follow the instructions for Gravity Bleeding, steps 6 and 7, to clean up and test the brakes.

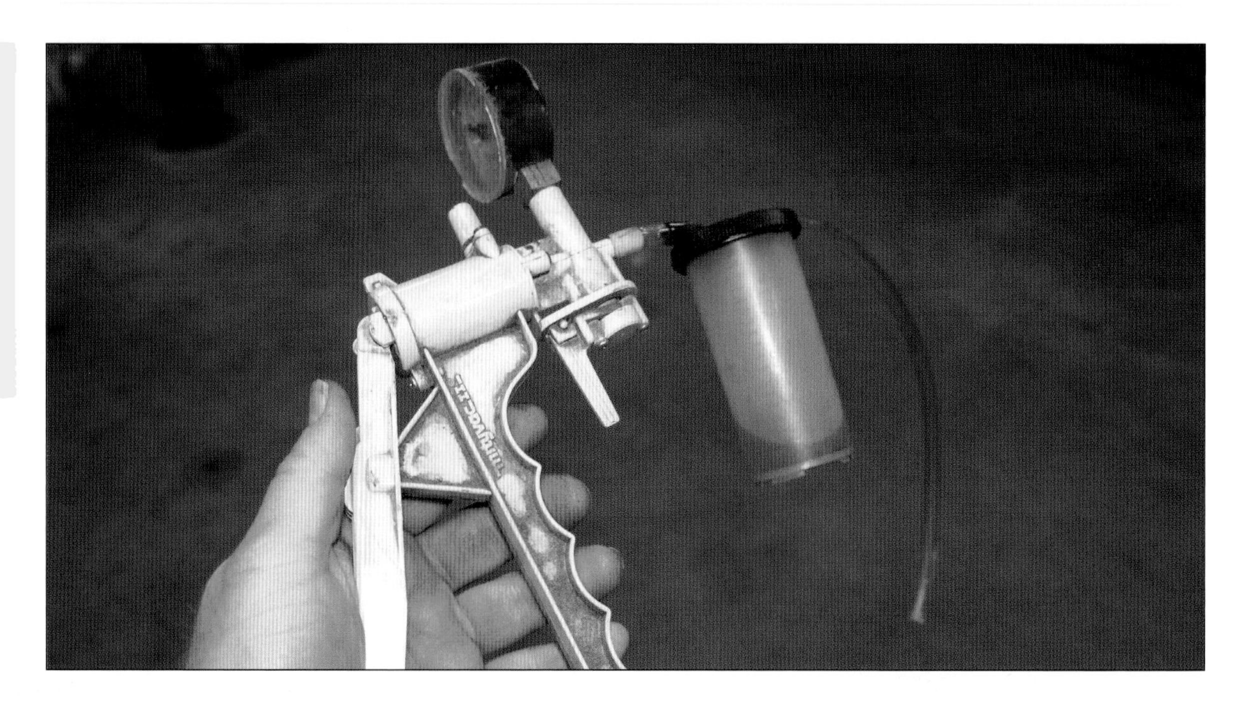

VACUUM BLEEDING

Vacuum bleeding brakes uses a vacuum pump to draw fluid out of the brake end of the system. It can be done by one person, but you will need to purchase a vacuum pump, which usually comes equipped with its own reservoir and tubing.

1. PREPARE THE VEHICLE FOR VACUUM BLEEDING.
   Follow steps 1 through 3 for Gravity Bleeding, and steps 2 and 3 for Pressure Bleeding.

2. ATTACH THE VACUUM PUMP TO THE BLEED SCREW.
   Attach the tube end of the pump to the bleed screw. Make sure the tube fits tightly around the screw to prevent air leaks.

3. APPLY VACUUM TO THE PUMP.
   Pump the vacuum pump up to about 15 psi and open the bleed screw. As the fluid begins to flow, check for bubbles through the clear tube. Keep about 10 psi on the line. When you don't see any more bubbles, close the bleed screws and release the vacuum.

4. REPEAT ON REMAINING BRAKE LINES.
   Repeat the process until all brakes have been bled. Be sure to check the fluid level in the master cylinder as you go.

5. CLEAN UP AND TEST THE BRAKES.
   Follow the instructions for Gravity Bleeding, steps 6 and 7, to clean up and test the brakes.

DISPOSING OF USED BRAKE FLUID

Used brake fluid is a hazardous material and must be disposed of properly. The rules for disposing of brake fluid and other automotive fluids vary by region, so check with your local recycling center or auto parts store to find out about the regulations in your area. In some places, you are allowed to pour the used brake fluid into a pan of kitty litter and let it evaporate. The litter can then be thrown away in non-recyclable waste.

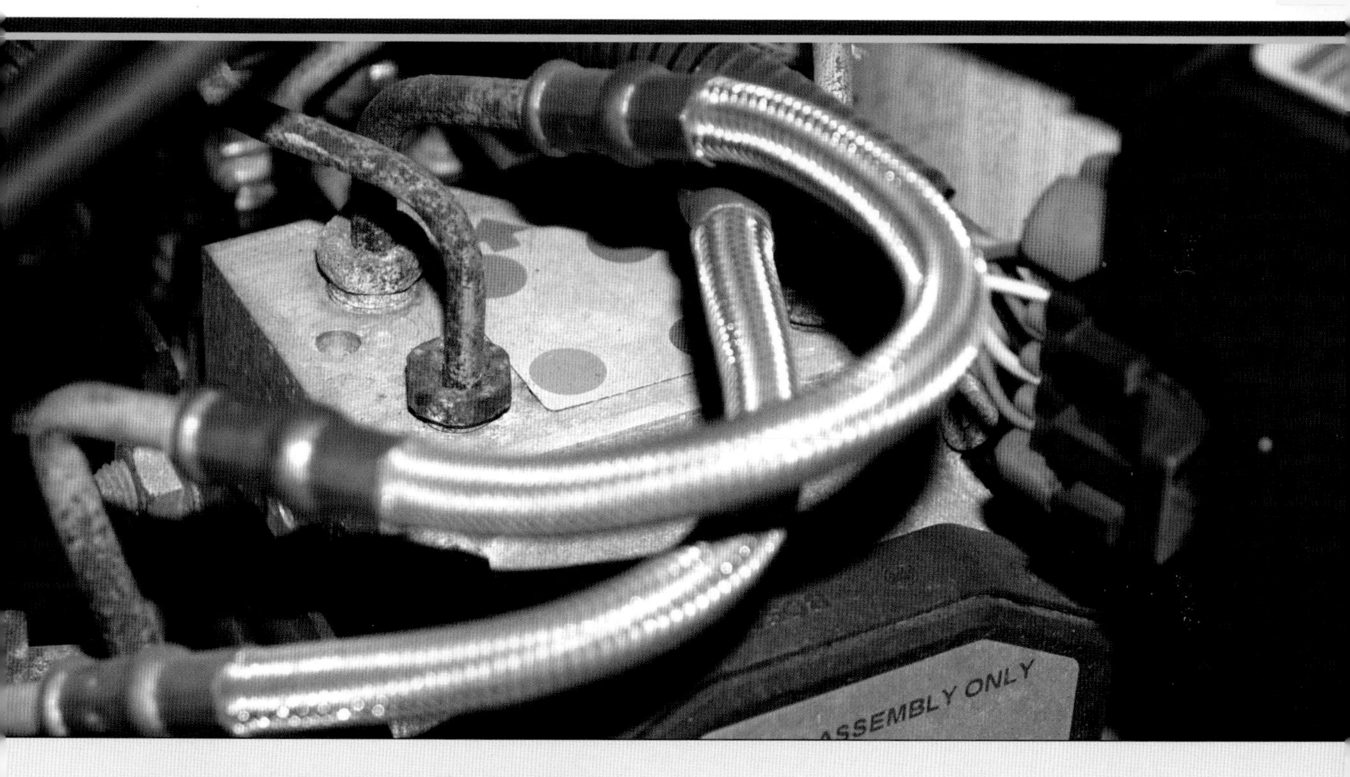

CARS WITH ANTI-LOCK BRAKES (ABS)

The anti-lock brake controller is located between the master cylinder and the brake ends. It has small passages to bypass fluid and pressure while operating. If air gets into the ABS unit, then it has to be purged by a professional using a pressure system and a computer. It's not possible to remove all of the air by simply bleeding the brakes alone.

If you have air in the lines after the ABS box, you can bleed the brakes normally. If you have air in the lines between the master cylinder and ABS box, you will need to have the ABS system purged by a professional. You can bleed the brakes as normal and then carefully take the car to the mechanic (without activating the ABS system). When in doubt, leave the brake bleeding to the professional.

STEERING AND SUSPENSION

IN THIS CHAPTER

How the Steering and Suspension Systems Work

Common Steering and Suspension Problems

How to: Check and Fill Power Steering Fluid

How to: Lubricate the Suspension System

How to: Inspect Shocks and Struts

How the Steering and Suspension Systems Work

Both the steering and suspension systems involve movement of the wheels, so they are usually talked about together. The steering, of course, allows you to point the car where you want it to go, and the suspension allows you to get there comfortably without feeling every bump in the road.

There are many different designs for suspension and steering systems. Here are some of the basic components used in most of them.

STEERING COMPONENTS

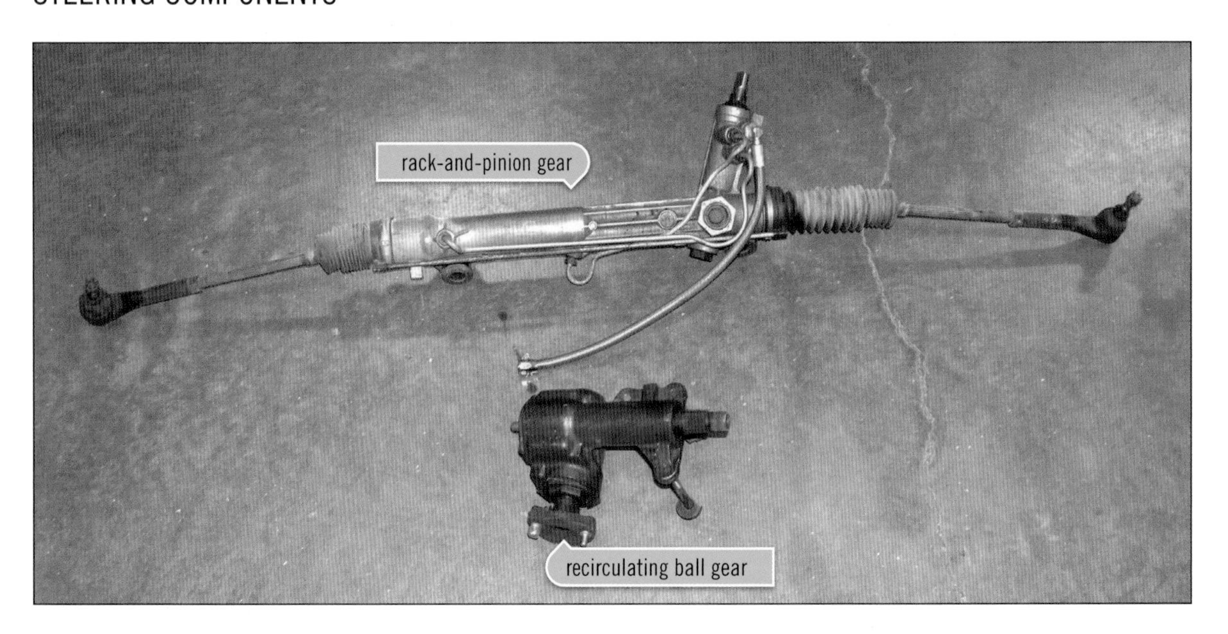

Steering Gears The steering gear converts the rotation from the steering wheel into the left and right movement that steers the front wheels (or all four wheels, in the case of four-wheel-drive vehicles).

There are two different steering gear styles: rack and pinion and recirculating ball. The rack-and-pinion gear has a long bar with gear teeth in it (the rack) that is run back and forth by a circular gear (the pinion). The recirculating ball gear also has a round gear in it, and it turns an outer arm back and forth. Rack-and-pinion gears are more popular in smaller vehicles, while the recirculating ball style is still used in larger vehicles.

Ball Joints, Tie Rods, and Linkage Ball joints and tie rods allow the steering gear to move while the suspension moves up and down. The ball joints work like a hip joint, with the ball rotating in a socket. This allows the suspension to move up and down and keep the tire flat to the ground.

A tie rod connects the steering gears to the drive wheels. It works just like a ball joint—it lets the connection between the steering and the spindle move around a central point. Most systems have four tie rods—two out by the wheels and two in by the inner rotation point, such as the ends of the rack. Some steering systems use connecting bars called *linkage* to hold and tie rods together and connect them to the steering gear.

Power Steering Power steering uses hydraulic or electric power to assist in the turning of the wheels. Hydraulic systems are run by a pump that is hooked to a belt that runs off the front of the engine. All-electric vehicles may use electric motors to assist in turning the wheels.

DRIVE-BY-WIRE (DBW) STEERING

Some new vehicles are steered by electric wires, not mechanical gears. Drive-by-wire steering senses when you are turning the steering wheel, then the computer sends a signal to actuators out on the wheels to move the wheels in the correct direction.

SUSPENSION COMPONENTS

Spindles The spindles hold the wheel and brakes in place, and rotate on the top and bottom mounts to turn the drive wheel. The spindle is usually connected to ball joints that can rotate and move with the suspension. This spindle is for a rear-wheel-drive car, and the shaft in the middle is connected to bearings that allow the wheel the spin freely. A front-wheel drive or four-wheel drive spindle will have a hole through which the drive spindle feeds and rotates the wheel.

The small arm on the spindle attaches to the steering tie rod, which allows the spindle to twist left or right, as needed.

Control Arms Control arms enable the spindles and or axles to rotate and give with the surface of the road. The spindle sits on the control arm, and the control arm is attached to the body. Most control arms use a ball joint to allow the joint to move as needed. The control arm can rotate up and down at the point where it is mounted to the body of the car. Depending on the design of the suspension, there may be one or two control arms per wheel.

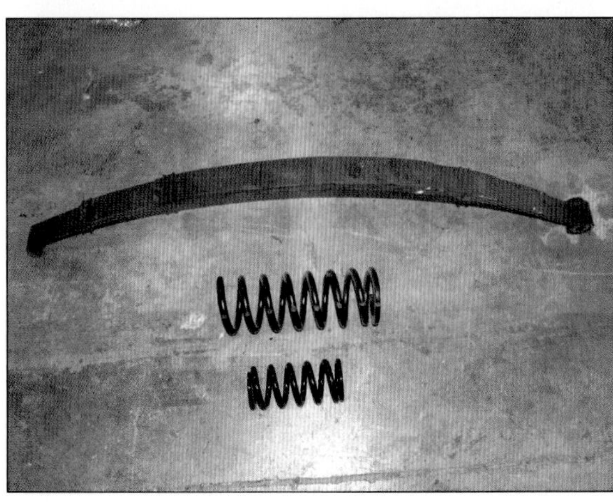

Springs Springs smooth your ride by enabling the wheels to move with road or driving conditions. They also help with turning by allowing the suspension to give a little bit. Springs absorb the impact from the road (potholes and uneven surfaces), but they aren't very good with dissipating energy. Left to themselves, they simply reverse the energy back through the spring as it rebounds. Some springs have a familiar coiled shape. Leaf springs are arc shaped.

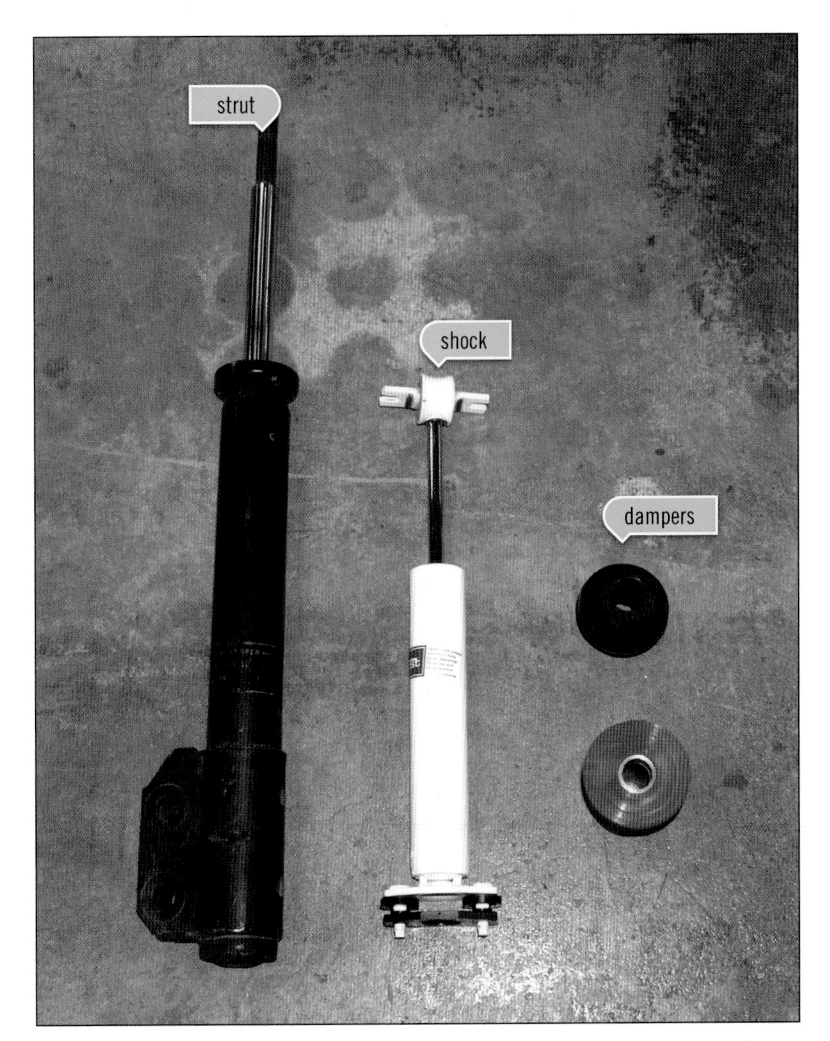

Shocks, Struts, and Dampers Springs are good at absorbing the impacts of the road, but they aren't good at dissipating energy. That's what shocks and struts do. They convert the energy of moving spring into heat, which can be dissipated. Shocks and struts use either gases or oil and valves to dissipate the energy pressing on the spring. When the spring wants to return to its original position, it slows the process down. This eliminates the "bouncing" effect of the springs and makes your ride smooth.

Struts (left) and shocks (center) work in the same fashion. Dampers (right) work like mini shocks—they take out the vibrations of the moving parts of the car and dissipate the energy to smooth out the ride.

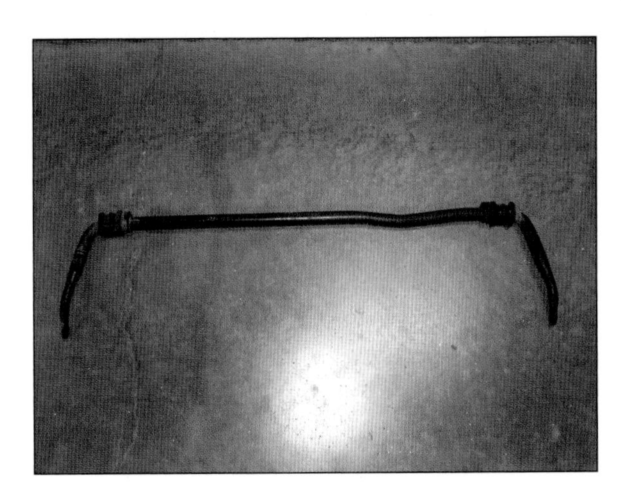

Sway Bars When your car turns, the forces acting on the car want to push the body outward, and this causes the body to want to "roll." The outside wheel is being pushed up and the inner wheel is being pushed down. Sway bars, sometimes called *anti-roll bars* or *stabilizer bars*, are used to stiffen the suspension when the car wants to roll. This enables car manufacturers to run softer springs (another way to reduce body roll) and improve ride quality. Sway bars are usually connected to the control arms and to the body by bushings that allow them to twist.

Common Steering and Suspension Problems

Steering and suspension problems can be difficult to identify because there are many issues that may or may not be tied to the suspension. A poorly wearing tire can be caused by problems with the suspension, steering, shocks, or may be caused by the tire itself. Suspension problems need to be addressed immediately to prevent an unsafe condition.

STEERING ISSUES

The car pulls to one side. If you feel the car pulling to one side as you drive, it could be due to any number of things: the front steering may need to be aligned, some of the steering components may have been damaged or bent, the power steering may have a blockage or a broken seal, or the tires may be different sizes or underinflated.

Check your tire air pressure and check for tire wear. If you don't see a tire pressure problem, have the car inspected and check the alignment.

The power steering fails. When the power steering fails, the car becomes very difficult to turn. This usually happens because the belt driving the power steering pump has broken. Another cause can be a leak in the hydraulic system.

If the belt fails, you may also lose the alternator and the charge light or indicator may show that the alternator is not working. If the belt is fine, look for signs of power steering fluid loss and check the power steering fluid level.

The steering becomes erratic. If it takes a lot of rotation of your steering wheel to get the car to turn, or the steering system "hangs up" in certain places, you may have worn steering components, like the tie rod ends or internal gears.

The power steering makes a whining noise. If you hear a whining sound from the power steering when you turn the wheel, it could be due to one of several issues: there may be air bubbles in the lines, the power steering pump may be failing, the belt driving the pump may be loose or failing, or a bearing on the pulley driving the pump may be failing.

Check the fluid level and belt condition first. If those are okay, the problem may be with the pump. Power steering pumps are not very efficient at forcing air bubbles out of the lines, and sometimes the air gets trapped in the system. This usually happens when the fluid level gets too low. Have a professional check to see if your system needs to be vacuumed out to remove the air from the lines.

SUSPENSION ISSUES

Vibrations Anything that moves in the front end can wear out and cause vibrations. The biggest culprits are usually the ball joints. As the ball joint wears it can't hold up against the pressure the tires exert.

Squeaks and Noises Loss of lubrication is the most common cause of squeaks. Sometimes you can eliminate a squeak by greasing the joints in the system. When rubber insulators wear out, they become brittle and the metal they contain will start to squeak. Springs and shocks can also be a source of squeaks and noises.

Poor Ride Quality If the car hits a large bump and the suspension bottoms out, the shocks may be wearing out or the suspension parts may be worn out and are not holding in place.

CHECKING FOR DAMAGE

If you suspect you have a problem with your suspension system, you can check for damage to specific parts.

Bent Components The tie rod on the top was bent in an accident; the one on the bottom is correct. Suspensions are generally symmetrical side to side, so check components against others to assess for damage.

Torn Boots This tie rod end boot was torn due to over-greasing the tie rod. Once the seal was broken, all the grease got pushed out and the cover tore itself apart. This tie rod will fail soon.

Rubber Components Check all the rubber components for dry rot, cracks, or splits.

Leaks Check the power steering connections for leaks, and check the grease fittings for grease coming out of the fittings.

Excess Grease Finally, check for globs of grease coming out of the boots and grease joints. Grease is a sign a joint has been overfilled, or that a cover or component may have failed.

GETTING AN ALIGNMENT

The front end of your car is designed to be aligned on a regular basis. Cars may need realignment due to normal wear or rough road conditions. If your car has an independent rear suspension, it also needs to be aligned.

Most new cars are designed to have all four wheels adjusted separately. Some older cars only require the front end to be aligned. Here are some of the terms your mechanic will use when talking about aligning your car.

Toe In and Toe Out If you look down at your feet, pointing your toes toward each other is "toe in," and pointing them away from each other is "toe out." On a car, you might think that you would want your tires to point straight ahead, but that isn't really true. On rear-wheel-drive cars, pointing the front wheels toward each other helps keep the car stable at high speeds. Front-wheel-drive cars are usually toe-out, since the front wheels are pulling the car rather than pushing. An all-wheel-drive car may have a combination of settings. Giving the tires a slight in or out position enables the car to drive straighter and handle turns better.

Camber Camber is a term that describes the angle of the tire in relationship to the road. When your tire tread is perfectly flat against the road, it has zero camber. If the top of the tire moves in toward the middle of the car, it has negative camber, and if it leans to the outside of the car, it has positive camber. Again, you might think that you would want the tire to be perfectly straight up and down, but most cars have some negative camber on the tires to provide better traction when turning corners.

Caster Caster is usually only adjustable on high-performance cars. Think of a bicycle front wheel and the fork that steers it. The fork is angled back toward the rider. This is called positive caster, and it adds stability to the bicycle at high speeds and helps stabilize the bike in turns. It works the same way on a car, and most cars have positive caster built into the suspension. Even if it isn't adjusted in your alignment, it will be on your spec sheet from the alignment shop.

How to: Check and Fill Power Steering Fluid

Most power steering systems use hydraulic power to help you push the wheels to the left and right. The power steering pump is usually driven off of a belt on the front of the engine. Check your fluid level and fluid condition on a regular basis and add fluid, if needed. The fluid level should always be checked when the engine is running.

WHAT YOU NEED

- Gloves and eye protection
- Paper towels
- Power steering fluid
- Funnel (optional)
- Turkey baster (if needed)

1. LOCATE THE POWER STEERING RESERVOIR. The reservoir may be part of the power steering pump, or it may be mounted remotely for ease of checking. Clean the top and cap of the reservoir to keep dirt and contaminants out of the tank while checking the level (FIGURE A).

2. TURN ON THE VEHICLE. Start the car to allow the pump to work fluid through the system. Turn the wheel to the left and right and listen for squealing or grinding noises. Loud noises may be a sign that the power steering pump is failing or that there is air in the lines.

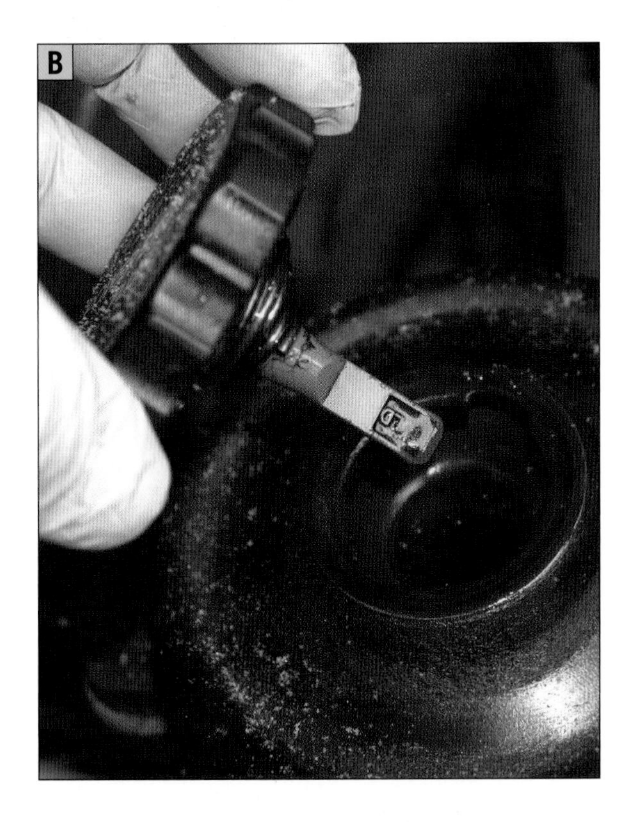

3. CHECK THE LEVEL OF THE POWER STEERING FLUID.
   Look at the side of the reservoir for marks indicating the proper level. If the reservoir is not translucent, there may be a dipstick in the cap, like this example, or the levels are marked inside the tank (FIGURE B).

4. CHECK THE CONDITION OF THE FLUID.
   Look into the reservoir and inspect the fluid. Depending on the manufacturer, the color of the fluid may be clear, red, or amber in color. Black contamination means rubber from the seals and hoses is seeping into the system, which can cause pump failure. If it is very dark and contaminated, have a professional check it out.

5. ADD FLUID AS NEEDED.
   If the level is low, add fluid. Use the recommended fluid for your car; generic power steering fluids may damage your vehicle. Be careful not to overfill, because hydraulic fluid expands when heated. Fill the reservoir a little at a time and recheck the level as you fill (FIGURE C).

6. REMOVE ANY EXCESS FLUID.
   If you find you have overfilled the reservoir, use a turkey baster to draw out some of the excess fluid.

7. REPLACE THE CAP.
   Replace the cap and wipe up any spills.

How to: Lubricate the Suspension System

Ball joints, tie rod ends, and other metal-on-metal parts need to be lubricated or they will grind against each other and fail. These parts have special grease fittings, sometimes called *zerk fittings*, that allow you to pump grease into the joint. Be careful when greasing the suspension; not enough grease and the parts are not protected, too much grease and you can break the rubber covers.

WHAT YOU NEED

- Grease recommended by manufacturer
- Grease gun
- Gloves and safety glasses
- Paper towels
- Jack and jack stands (if needed)

A

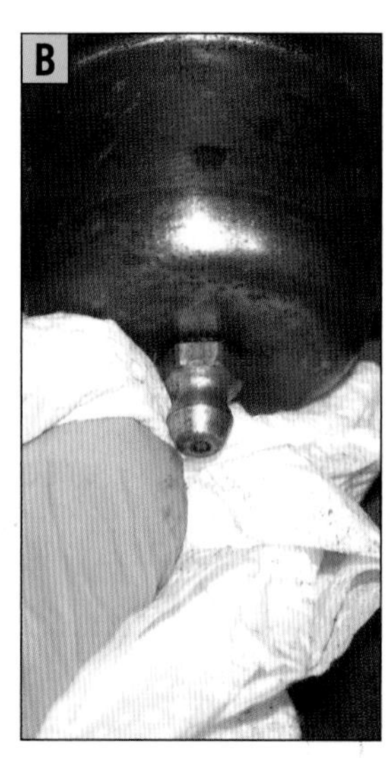
B

1. LOCATE THE GREASE FITTINGS.
Check the suspension parts for grease fittings. These look like small bolts with rounded tops (FIGURE A). You may need to raise the vehicle and secure it with safety jacks to access the suspension system. Locate all of the grease fittings on the car.

2. CLEAN THE FITTING.
Use a paper towel or clean rag to wipe off the fitting prior to filling. This will prevent dirt from getting into the joint (FIGURE B).

USING A GREASE GUN

Most grease guns use a crank handle to force grease into the joints. How they work is simple: the grease cartridge inserts into the tube, and the piston on the end drives the grease down as it is pumped.

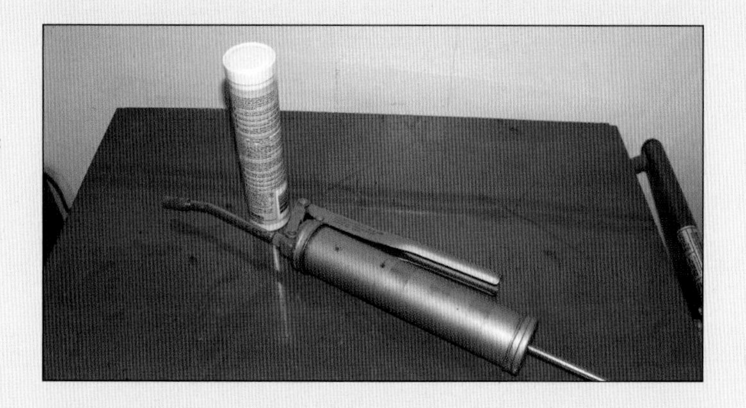

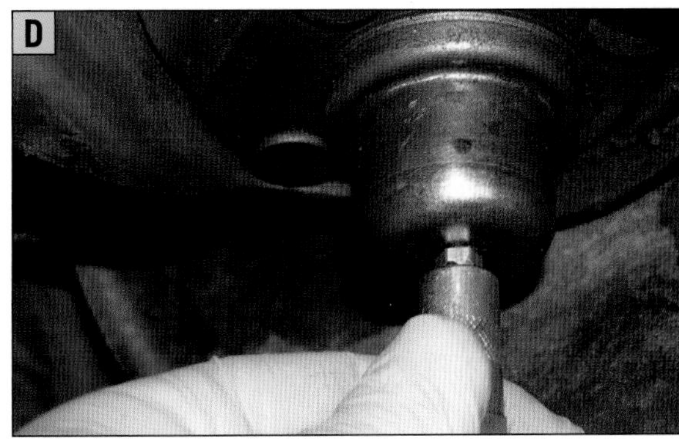

3. **CLEAN THE GREASE GUN NOZZLE.**
   Clean the old grease and dirt off of the nozzle so you don't inject dirt into the joint (FIGURE C).

4. **INSTALL THE GUN NOZZLE OVER THE FITTING.**
   Press the grease gun nozzle firmly onto the fitting and hold it tight while pumping new grease into the joint (FIGURE D).

5. **FILL THE JOINT.**
   Fill the joint until you see the top side boot bulge slightly outward (FIGURE E). Installing too much grease can cause the cover to rupture. Once you have installed the grease, proceed to the next joint and fill until all the joints have new grease.

FILLING JOINTS WITH RELIEF VALVES

Some ball joints and tie rods have a relief valve or outlet on them, usually at the top of the boot cover or opposite the end of the grease fitting. The purpose of the relief valve is to allow the old, contaminated grease to flow out of the joint and be replaced with fresh grease, much like a U-joint. However, not all joints have this feature, and you can cause damage to the joint by overfilling them and having the grease come out the side of the cover.

If your suspension joints have a relief valve, you should see grease come out the top of the boot or opposite the grease fitting, and it should come out fairly easily. If it comes out the sides of the boot, or the boot swells quite a bit, stop. When a boot ruptures or breaks its seal, dirt and contaminants can get into the joint.

How to: Inspect Shocks and Struts

Shocks and struts wear out gradually, and you begin to adjust your driving habits to wear without even realizing it. Worn shocks or struts can also affect the comfort of your ride and increase the stopping distance of your car, which can be hazardous. Check your shocks every year, or every 12,000 miles (19,000 km).

There are three ways you can check your shocks. The best way is to simply drive the car and pay attention to how the car performs. A visual inspection and an old-fashioned "bounce test" can also provide information about the condition of your shocks and struts.

CHECK FOR PROBLEMS WHILE DRIVING

Monitoring your car's performance while driving is the best way to check your shocks and struts for wear. Find a safe place, such as an empty parking lot, to see how the car responds to hard stops and bumps. These are some things to look for when driving.

The car does a "nose dive" when brakes are pressed. If the nose of your car dips toward the road under hard braking and bounces up when you come to a stop, your shocks are probably worn out.

The car leans when turning corners and in high winds. When you take a hard corner, the body of the car wants to roll to the outside. The shocks are supposed to help keep it level. If your car is leaning noticeably to the side when turning, the shocks are worn. Also pay attention to your car's movement when a big truck passes on the highway. If the rush of wind causes your car to lean significantly, it's another sign that your shocks may need attention.

The car bottoms out when you hit a bump. Another sign of worn shocks is if the suspension bottoms out when you hit a bump (you'll hear a thunk and feel a harsh jolt) or if you notice the car springing up and down like a boat on rough water.

You feel a strong vibration in the car or steering wheel. Modern shocks use several different valves to work—some for low rumbles from the road and some for big potholes and heavy turns. If you are really feeling the rumble of the road through the steering wheel, or the whole car feels like it is vibrating, it may be coming from the shocks.

The suspension travel is limited. If your shocks have been damaged, they may not be able to move at their full length, and you will get a hard stop from the damage. This can cause a hard "thunk," or the whole body may move. When this happens, the shock will destroy itself internally very quickly.

PERFORM A VISUAL INSPECTION

You may need to raise your car to get a clear view of the shocks. Here's what to look for.

Oil Leakage Oil can leak from the seals around the piston or from external damage to the shock. Regular road grime is normal, but if it is mixed with oil from the shock, the shock is failing and needs to be replaced.

Shock Damage If the tube body of the shock becomes dented, it can prevent the oil, valves, and pistons inside from moving properly. Inspect the exposed piston for signs of rust, scratches, or contamination—these can rub against the seals and cause seal failure.

Tire Cupping or Uneven Wear If a shock is not working properly, it may allow the tire to bounce up and down, causing uneven wear. This uneven wear is called "cupping." Uneven wear can also be caused by misalignment.

USE THE "BOUNCE" TEST

The bounce test is simple: push down on one corner of the car. When you release the pressure, the corner should come up to its original position without bouncing. If it does bounce, the shock is probably worn out and needs to be replaced by a professional.

TIRES AND WHEELS

IN THIS CHAPTER

Wheels 101

The wheels on a car are generally trouble-free, but there are a few things you should know about them. Wheels are usually made from steel or aluminum. They are designated by the rim size, the number of holes used to hold them on the vehicle, and by the backspace and offset measurements. Make sure you use wheels designed for your vehicle.

The rim of the wheel is a vulnerable spot and can be bent if the wheel hits a pothole or curb. A bent rim can create imbalance in the tire and may also cause vibration or prevent the tire from sealing properly. Minor bends can usually be straightened, but a severely bent rim will need to be replaced. Straightening a wheel should be done by a professional.

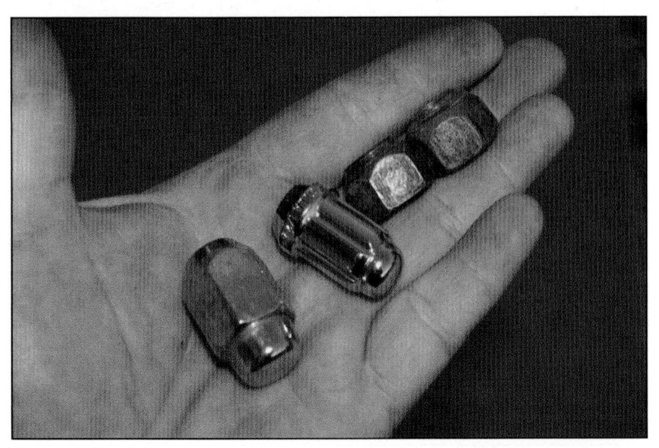

Lug nuts are used to hold the wheel onto the car. Some are rounded, and some are tapered at the point where they meet the wheel. Make sure you are using the correct lug nuts for your vehicle, and do not mix or change types of lug nuts; it is dangerous and can cause damage.

Lug nuts are designed to use a specific torque, or correct tightness. If you have your tires rotated or changed by a professional, they should set this properly. If you rotate your tires yourself or if you have an emergency flat tire, make sure you use the correct torque specification for your wheels. This should be in your owner's manual. If you don't have a torque wrench, take the car to a professional to have it checked.

OFFSET AND BACKSPACE

Offset and backspace are two different ways to determine the location of the wheel when it is bolted to the hub of the car.

Offset is the distance between the mounting surface and the centerline of the wheel, and it is measured in millimeters. Offset can be positive, negative, or zero. If the mounting surface is closest to the street side of a wheel, the offset is positive. If it is closest to the brake side of the wheel, it is negative. If it is aligned with the center of the wheel, the offset is zero.

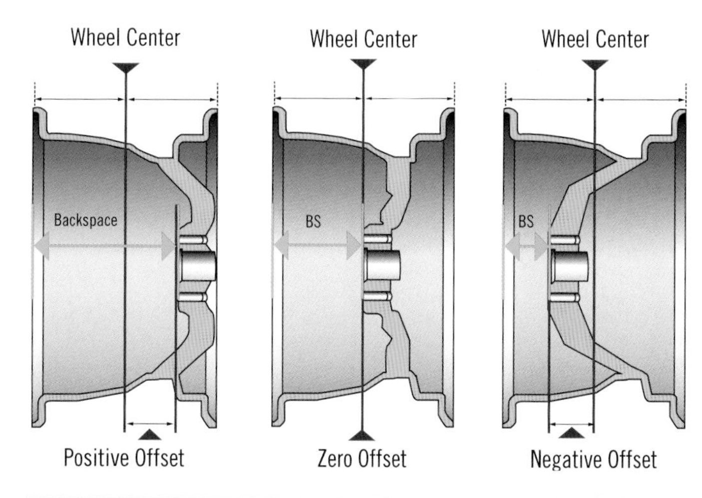

Backspace is the distance between the mounting surface and the back edge of the wheel (the edge closest to the brakes). It is measured in inches.

For example, a 10-inch wheel that is mounted right in the middle has a backspace of five inches and zero offset. A six-inch backspace means the wheel is closer to the outside and has a positive offset. A backspace of four inches means the wheel is mounted closer to the brakes and has a negative offset.

Backspace is important to consider when choosing wheels. Too much or too little can affect the car's performance, and the wheels can touch or rub other parts and cause damage. Front- and rear-wheel-drive cars generally have different backspacing; don't use wheels intended for a different type of car.

Tires 101

You may not realize it, but your tires are probably the most neglected part of your car. They are also the one area that, if maintained properly, can affect your pocketbook the most. Proper tire maintenance can improve your gas mileage, lengthen the lifespan of your tires, and most importantly, keep you safe.

The most important tool to carry in your car may be the **tire pressure gauge**. There are a few different styles of gauges, including dial gauges and stem-type gauges (shown). Checking your tire pressure regularly and adding air when needed is an easy way to save money and additional costs down the road.

The **valve stem** is where you add air to your tires. They can be made from rubber or a combination of metal and rubber. Leaks can occur at the seal between the stem and the valve.

TIRE PRESSURE MONITORING SYSTEMS

Since 2008, all vehicles in the United States have been required to have a **pressure monitor** mounted inside the tires. These monitors measure the air pressure in the tire and send a signal to the car's computer if they detect a problem. The computer then activates a warning light on the dashboard. If the light comes on, inspect the tires and check the air pressure. Sometimes the monitors can go bad, and if they do, they will need to be replaced.

TYPES OF TIRES

In the old days, tires were made for specific driving conditions: dry, rain, mud, or snow. Most tires today feature an all-season design, which is probably all you need. However, you can still buy specialty tires to suit your driving needs. Tire design can affect the handling, performance, and stopping distance of your vehicle.

All-Season Passenger Tires Most passenger cars use a tire that has been designed to work in any weather condition. The ride quality, ability to drive in certain conditions, and durability will depend on each manufacturer's design.

Performance Tires Performance tires are made for cars that can travel at higher speeds. They are designed to have more rubber in contact with the road, and are made of softer rubber that grips the road better than regular tires. These tires are not as good as all-season radials in bad weather or adverse road conditions, and the softer rubber will wear away faster.

Truck Tires Because truck tires need to carry heavy loads, they are generally thicker and heavier than regular passenger tires and have large treads for handling mud and gravel. They are not usually as quiet as passenger car tires.

All-Terrain Tires These extreme truck tires are designed for off-road driving. They are very thick and heavy, and are designed to handle irregular off-road conditions.

Snow Tires If you live in an area that gets a lot of snow, you might consider having a set of snow tires. These tires have thicker treads than regular tires, and may even have studs that screw into the tire for better traction in very heavy snow conditions.

Choosing the Right Tires

The first thing to check when choosing tires for your vehicle is the tire and loading information sticker, which is usually on the inside edge of the driver's side door frame. This sticker provides the specifications for the type of tire that is best for your vehicle. For optimal performance, always install the type of tire recommended by the manufacturer. Changing the tire type, tire size, load rating, or speed rating can dramatically affect your car's performance.

READING THE TIRE STICKER AND TIRES

The tire sticker will have size information for the front, rear, and spare tires, and may have the load rating and speed rating as well. This information is coded using a series of letters and numbers. These codes may vary somewhat by manufacturer.

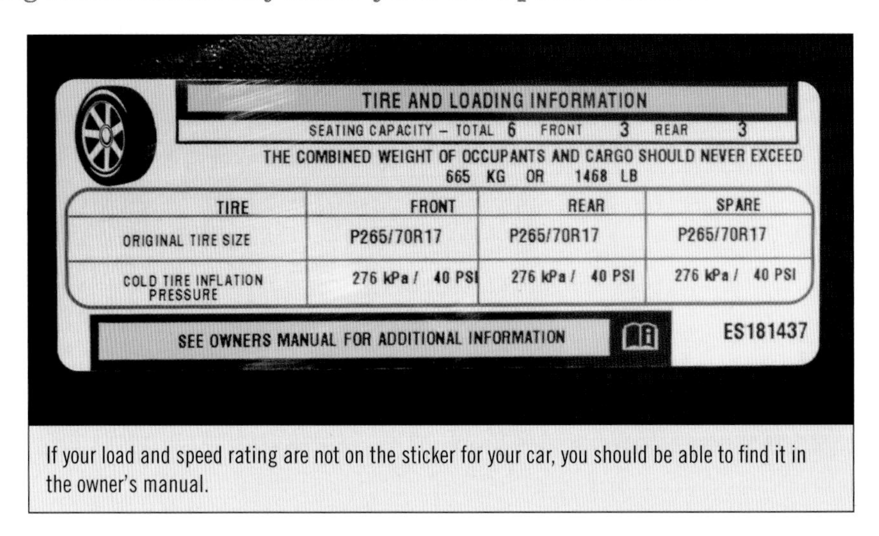

If your load and speed rating are not on the sticker for your car, you should be able to find it in the owner's manual.

LOAD RATING

The load rating is a number from 0 to 279, and it can usually be found on the tire sticker or in the owner's manual, along with the tire size information. The load rating is very important: it tells you how much weight the tire is designed to carry. The higher the number, the greater the load the tire can carry. Always use tires designed for the load rating of your vehicle. Using a tire with too low of a rating can result in a burst tire, and using a tire with too high of a rating can adversely affect the handling. The load rating of the tire is directly related to the weight of the car and the maximum weight your car or truck can carry.

SPEED RATING

The speed rating is a letter between A and Z that tells how fast you can safely drive on the tires. In general, the later the letter falls in the alphabet, the higher the speed at which the tire can safely operate.

Most off-road truck tires are rated "L" (75 mph/120 kph), typical passenger tires are rated around "S" (112 mph/180 kph), and high-performance tires may be rated "Z" (149 mph/240 kph) or up to "(Y)" (over 186 mph/over 300 kph). When choosing a speed rating, make sure you select a tire that meets or exceeds the type of driving you will be doing.

The same codes that you see on the tire sticker can be found on the edge of the tire. Let's take a look at what each part of the code stands for. (Keep in mind that your vehicle may have different numbers.)

1. **P** or **LT**: This tells you what kind of tire it is. "P" stands for passenger tire; "LT" stands for light truck. Choose a tire based on your vehicle type.
2. **195**: The number following the "P" or "T" tells how wide the tire is in millimeters.
3. **75**: This number is the aspect ratio. It means the tire is about 75 percent as tall as it is wide. Performance tires generally have a low aspect ratio to eliminate sidewall flex at high speeds, and truck tires may have taller aspect ratios.
4. **R**: This stands for "radial construction." Sometimes it is accompanied by a letter that denotes the speed rating for the tire. Most tires, with the exception of some trailer or specialty tires, are radials.
5. **14**: This is the size of the rim in inches.
6. **92S**: This is one way to display the load rating and speed rating of a tire. 92 is the load rating, and S is the speed rating. Some tires have the speed rating matched with the load rating, and some pair it with the "R" for radial design.
7. **M+S:** This designates the tire as useful in mud or snow conditions, and you will see it on most all-season radials.

BUYING USED TIRES

As a tire ages, the rubber begins to dry out and the tire loses its ability to hold air, which can cause blowouts and failures. Tires don't need to be used to deteriorate; even on a new tire, the rubber can dry out if it's just been sitting in a tire shop for years. If you're considering purchasing used tires, it's important to look for signs of aging, such as dry rot and blisters, as well as wear.

Common Tire Problems

Get in the habit of taking a daily walk around your car to check the tires. This can help you locate problems early on and prevent them from getting worse. Here are some of the more common tire problems and what to do about them.

The tread is worn on both sides, but not in the middle. This problem is caused by an underinflated tire. Check for leaks in the tire and valve stem, and add air pressure to the manufacturer's recommendation.

The center of the tread is worn, but the sides are not. This is caused by overinflating the tire. Check the pressure in the tire and reduce to the manufacturer's recommendation.

One side of the tread is worn more than the other. This is usually an indication of an alignment problem. Have your car inspected for proper alignment.

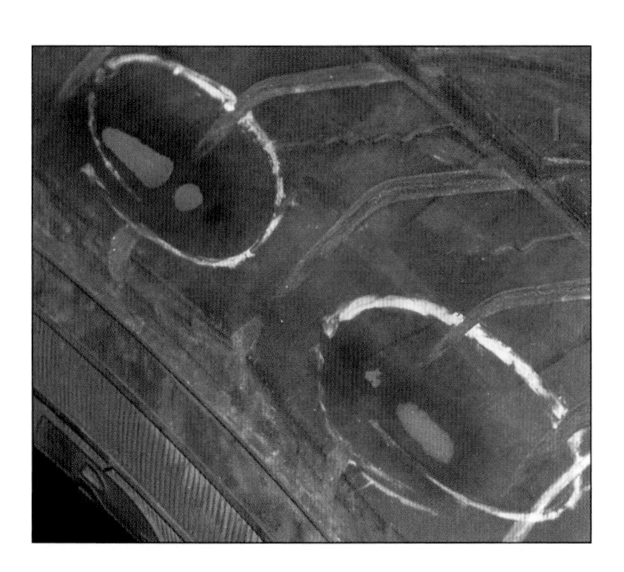

The tire shows "cupping." Cupping looks like a series of peaks and valleys or a scalloped pattern. It can be caused by several things, but the most common cause is worn suspension parts that allow the tire to bounce. If the cupping is due to worn suspension components, then the parts will need to be replaced. The rear tires of a front-wheel-drive car may also show cupping, since the weight of the car is over the front wheels making it easier for the rear of the car to bounce. Proper tire rotation helps with this wear.

The tire looks underinflated. A low tire can be caused by normal use, or it may be due to a puncture or improper seal of the tire and the rim. If you notice a low tire, check for objects like nails poking into the tread, and listen for air escaping the tire. If you don't see a problem, try adding air to the tire and re-checking the tire pressure. Do not drive on a tire that is severely low. If you do not have air available, you should change the tire as quickly as possible. If the tire is leaking or you have found a puncture, change the tire and take it to be repaired.

The tire has blisters and bubbles. Blisters and bubbles indicate trouble. If the blister is on the side of the tire, the tire was probably pinched when it hit a pothole, forcing the sidewall between the hole and the rim. If the blister is in the tread area, it means the reinforcement cords in the tire have failed. Blisters can cause blowouts—replace the tire immediately.

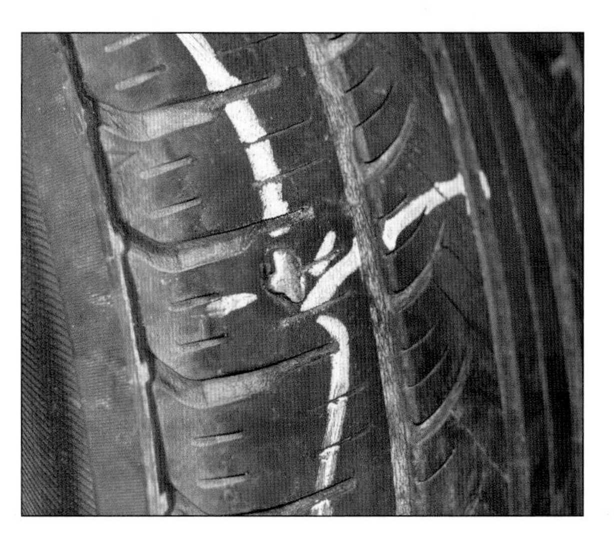

The tire has a puncture. Tires can become punctured in many ways, but often it's due to a foreign object picked up from the road, such as a nail or screw. If the puncture is in the tread area, it can be repaired fairly inexpensively. Sidewall punctures cannot be repaired and the tire will need to be replaced.

How to: Check the Tire Pressure

Maintaining proper air pressure in your tires maximizes the life of the tires, saves money on gas, and helps the car handle better while driving. It is recommended that you check your tires cold, because the air inside can expand as the tire heats up from road friction. However, if you see a problem, go ahead and check them warm. Tire pressure is measured in PSI (pounds per square inch). Check your owner's manual or the sticker on the driver's side door for the correct PSI for your specific vehicle.

WHAT YOU NEED

- Tire pressure gauge

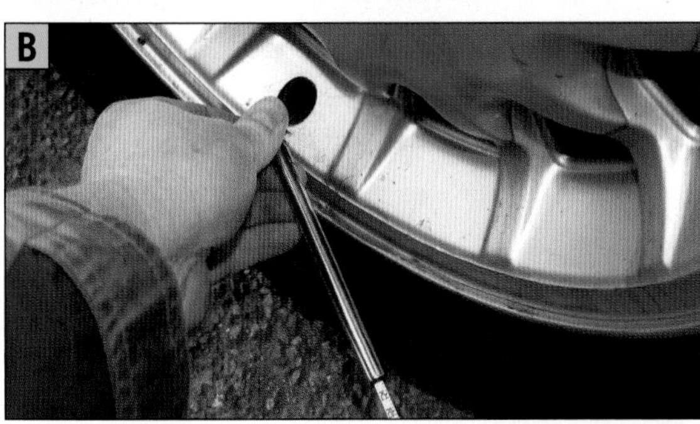

C

FORD / MERCURY

RECOMMENDED TIRE SIZE and INFLATION PRESSURE (COLD)
DIMENSIONS DES PNEUS et PRESSIONS DE GONFLAGE RECOMMANDÉES (À FROID)

TIRE SIZE / DIMENSIONS DES PNEUS	TIRE PRESSURE FRONT		AVANT	PRESSION DES PNEUS REAR		ARRIÉRE
P215/70R15 97S*	220 kPa	32lb/po²	32 PSI	240 kPa	35lb/po²	35 PSI
P225/70R15 100S* P225/60R16 97T*	220 kPa	32lb/po²	32 PSI	240 kPa	35lb/po²	35 PSI
T125/80R16	415 kPa	60lb/po²	60 PSI	415 kPa	60lb/po²	60 PSI
P225/70R15 100V*	240 kPa	35lb/po²	35 PSI	240 kPa	35lb/po²	35 PSI

* MUST BE REPLACED WITH AN EQUIVALENT TYPE SPEED RATED TIRE.
* NE REMPLACER QUE PAR UN PNEU DONT L'INDICE DE VITESSE EST LE MÉME.

TOTAL LOAD = OCCUPANTS PLUS LUGGAGE CHARGE GLOBALE = OCCUPANTS PLUS BAGAGES

MAXIMUM LOAD / CHARGE MAXIMALE	OCCUPANTS / OCCUPANTS	DISTRIBUTION RÉPARTITION FRONT AVANT	REAR ARRIÉRE	LUGGAGE BAGAGES
499 kg/1100 lb	6	3	3	91 kg/200 lb
431 kg/956 lb POLICE ONLY / POLICE SEULEMENT	5	2	3	91 kg/200 lb

FOR SUSTAINED HIGH SPEED, TRAILER TOWING, RECREATIONAL ACCESSORIES AND TEMPORAL SPARE USAGE – SEE OWNER GUIDE.
VITESSES SOUTENUES, REMORQUES, ACCESSOIRES DE PLAISANCE ET PNEU DE SECOURS PROVISOIRE:
F7AC-1532-A

1. LOCATE THE VALVE STEM.
   The valve stem should be sticking out of the wheel edge (FIGURE A). Unscrew the valve stem cap from the valve stem. If the stem does not have a cap, you should purchase a replacement. The cap helps protect the valve from dirt and debris.

2. GET YOUR READING.
   Press the end of the pressure gauge to the top of the valve stem, making sure that air does not escape (FIGURE B). On a post-style gauge, the post will extend and stay at the maximum pressure point. On a dial gauge, or a gauge mounted on the end of a compressor, the pressure may need to be read while the gauge is still attached to the stem.

3. COMPARE THE READING.
   The correct pressure for your specific car is printed on the tire sticker or in the owner's manual (FIGURE C). Compare your reading and adjust the air pressure up or down to match the recommended pressure. The reading should be within a pound of pressure one side or the other. Replace the valve cap.

How to: Add Air to a Tire

If your tire needs air, you will need access to an air compressor to fill the tires properly. Most service stations have air available.

WHAT YOU NEED

- Tire pressure gauge
- Air compressor (with air chuck)

1. **PRESS THE AIR CHUCK ONTO THE VALVE STEM.**
   Hold the air chuck onto the stem and keep it square so air does not escape around the valve. Listen for the sound of air rushing into the tire. Add air in small increments, checking the pressure as you fill (FIGURE A).

2. **RELEASE EXCESS PRESSURE IF NEEDED.**
   If you overfill the tire, you can let some air out using the little post on the back of your pressure gauge. Press the post to the center pin of the valve. Remember to go slowly and check often. Once the air pressure is set to the manufacturer's recommendation you can replace the cap (FIGURE B).

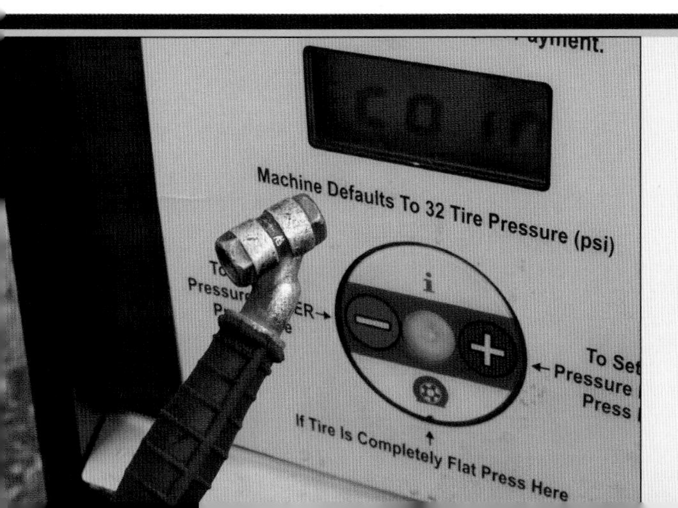

AIR CHUCK

The air chuck looks similar to a tire pressure gauge and is hooked to the end of the compressor. Air compressors at service stations should have one built into the air line. Some, like this one, have a built-in regulator so you don't over fill. If you have a home compressor, you will need to install an air chuck to the compressor hose.

How to: Rotate Your Tires

Since the front tires move from side to side while the vehicle is running, they become scuffed more quickly than the rear tires. Rotating your tires helps them wear evenly and improves the performance balance of the car. Some tire manufacturers require regular rotation for their warranties to remain valid.

- Jack and jack stands
- Wheel chocks
- Lug wrench
- Torque wrench
- Socket for lug nuts
- Locking lug nut key (if needed)
- Tools to remove decorative covers (if needed)

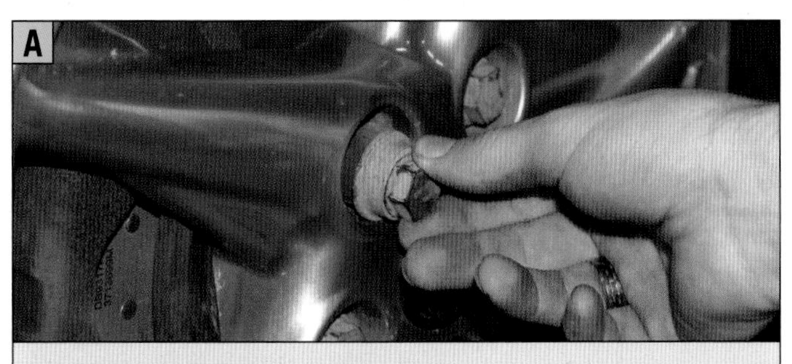

This car has plastic covers for the lug nuts.

1. **PREPARE THE VEHICLE AND TIRE.**
   Place the car in park or in gear and place wheel chocks around a wheel that isn't being rotated. Remove any decorative wheel covers to expose the wheel lug nuts (FIGURE A).

2. **LOOSEN THE LUG NUTS.**
   While the car is on the ground, use a socket and wrench or a lug wrench to slightly loosen the lug nuts, but don't pull them away from the wheel (FIGURE B). If you try to do this after you get the car in the air, the wheel will spin.

3. **RAISE THE VEHICLE.**
   Follow the instructions in this book or your owner's manual for safely raising your vehicle using a jack (FIGURE C).

If you choose to rotate your own tires, make sure you have the proper tools and safety equipment before undertaking the job.

ROTATION PATTERNS

The way you rotate your tires will depend on the layout of your car (front-wheel drive vs. rear-wheel drive) and the type of tires you have. Your owner's manual should show the proper rotation procedure for your vehicle. If you don't have the manual, follow the diagrams shown here.

For a Full-Size Spare: If your car has a full-size matching spare, you can mix it into the rotation. Don't rotate the spare if it is a temporary spare, a different size, a different type wheel (aluminum vs. steel), or different brand.

For Directional Tires: If your car has directional tires, which are designed to be used on one side of the vehicle only, your rotation pattern is limited to swapping the fronts to the rear on the same side (1 to 3 and 2 to 4). The tire will usually have a notation on the side showing the rotation direction.

For Differently Sized Tires: If your car has bigger tires and rims on the back than on the front, then your rotation is from side to side on that axle only (1 to 2 and 3 to 4).

Front-Wheel-Drive Vehicle

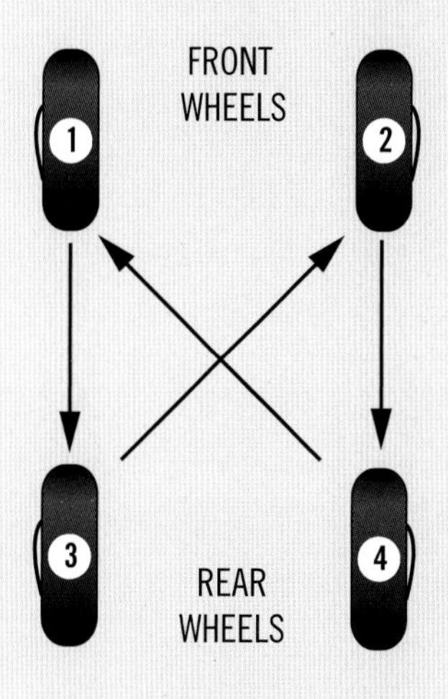

Rear-Wheel-Drive Vehicle

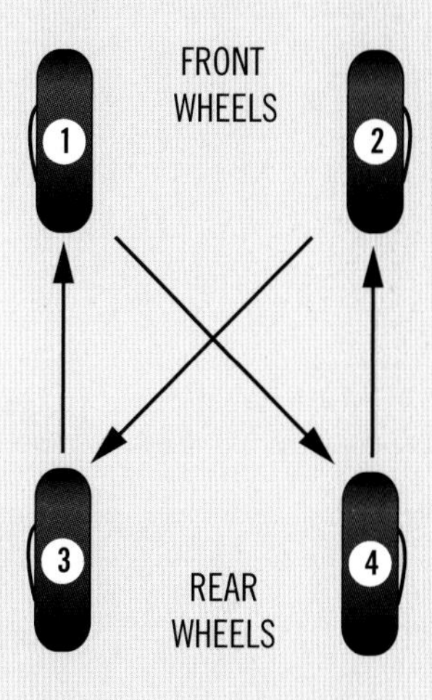

Front-Wheel-Drive Vehicle with Full-Size Spare

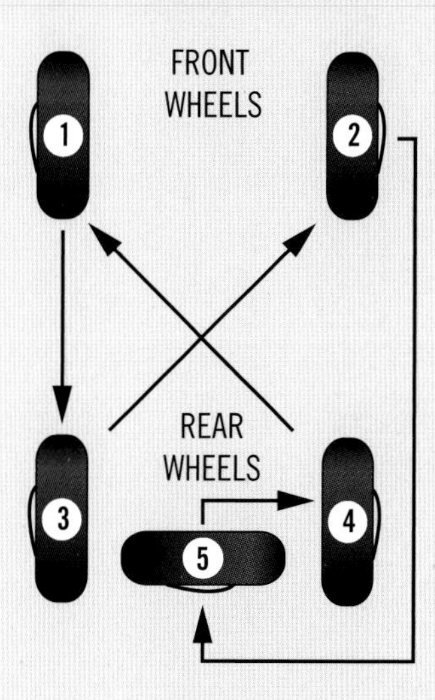

Rear-Wheel-Drive Vehicle with Full-Size Spare

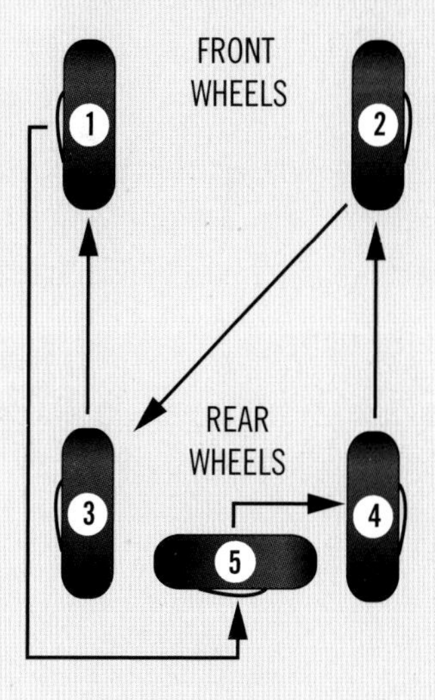

D

E

F

4. PLACE THE JACK STANDS.

Once the wheel is off the ground, place a jack stand in a position that will be safe, such as underneath part of the frame. Use the car jack to slowly lower the car onto the stand until the weight of the car is supported by the jack stand (FIGURE D).

Once the car is secure, lift the second tire off the ground and secure it with another jack stand. Give the body of the car a gentle push to make sure it's stable.

5. PULL THE WHEELS.

Remove the lug nuts and pull the wheel off of the body. Be careful and use proper lifting procedures—wheels and tires can be heavy. Pull the next wheel in the rotation so the previous wheel can be mounted (FIGURE E).

6. INSPECT THE VEHICLE.

On a front-wheel-drive car, the left front tire is rotated to the left rear (FIGURE F). While you have the wheels off, take a minute to inspect the brakes and suspension components for damage or wear.

G

H

7. INSTALL THE FIRST WHEEL.
Place the wheel in its new location (FIGURE G). Replace the lug nuts, tighten them by hand, and snug with your wrench. Use the car jack to lift the car off of the jack stand supporting this wheel. Remove the jack stand and lower this wheel to the ground.

8. TORQUE THE LUG NUTS.
Once the tire is touching the ground, you can torque the lug nuts (FIGURE H). Always torque the lug nuts in a star pattern, not a circular pattern. Check your manual for the correct torque specification for your vehicle. Replace the wheel covers, protective covers, or lug covers.

9. REPEAT WITH REMAINING WHEELS.
Repeat the process with the remaining wheels until all wheels have been rotated.

THE EXTERIOR

IN THIS CHAPTER

Common Exterior Problems

Washing and Waxing Your Vehicle

How to: Wash a Car

How to: Wax a Car

Paint Imperfections and Rust Problems

How to: Remove Scuffs in Paint

How to: Restore Headlight Lenses

How to: Inspect and Replace Wiper Blades

How to: Inspect and Repair Weather Stripping

Common Exterior Problems

The exterior of your car is subject to a wide range of environmental conditions and situations that can cause wear or damage. It's vulnerable to everything from the salt used on icy roads to the careless person in a parking lot who hits your car with a shopping cart. Most of these hazards are hard to avoid, so you need to know how to address damage when it occurs. Here are the primary components of the exterior and some problems that can affect them.

PAINT

The paint does more than just keep your car pretty—it protects the metal surfaces underneath from rust. The paint seals the metal and prevents oxygen from causing corrosion. Paint is exposed to all kinds of hazards—road debris, insects, acid rain, UV rays, road salt, and humidity. If the paint becomes damaged, the metal underneath can be exposed and the corrosion process begins. The best defense against this is to keep your paint clean and waxed, and fix chips or other damage as soon as possible.

If your paint is fairly new and it begins to fade, peel, or bubble, check with your manufacturer to see if there has been a paint recall. Paint technology is frequently changing to make it less toxic, but not all formulas perform as well as expected.

GLASS

The front windshield, side windows, and rear windshield are all made of glass, but they perform differently. The windshield is made of laminate glass, which has a plastic layer to prevent the glass from shattering and spraying everywhere on impact. It also cushions an impact should you come in contact with it. You may be able to have small cracks and chips in laminate glass repaired, but the longer you leave a crack or chip, the greater chance it will spread, requiring a windshield replacement. Cold weather can also accelerate an expanding crack.

Scratches in windshields are very difficult to remove. Most scratches occur when the wiper blades wear out or are ripped, and the metal arm of the wiper blade scratches the windshield. Polishing scratches out of a windshield is hard, and the products designed to remove them are often temporary or ineffective, so it's best to have a damaged windshield replaced.

The side windows and rear windshield are usually made from tempered glass, which shatters into very small pieces when it breaks. This reduces the chances of injury from a long, sharp piece of glass. If tempered glass shatters, it must be replaced.

BODY DAMAGE AND RUST

Accidents happen, and when your car gets a ding or a crunch, the paint can chip, exposing the body panels to corrosion. Most newer cars have panels made from steel that has extra galvanizing , which reduces rust, but some cheaper replacement panels may rust much faster because of the quality of the metal used.

Some forms of damage, like dents from hail, may be candidates for paintless dent repair (PDL). If the metal hasn't been stretched or bent too much, and the paint itself hasn't been lifted or damaged, a PDL professional can sometimes gently push the metal back in place. PDL can be much less expensive than conventional paint repairs.

Rust problems can't be covered up; they need to be repaired or replaced. There are some temporary fixes to keep the problem from getting worse, but the only way to properly repair rust is to remove it.

HEADLIGHTS

The clear plastic housing that covers the headlight bulb can become cloudy and dull over time. Road debris, impacts, and UV rays degrade the clear plastic, causing the headlights to dim and become less safe and effective. When this happens, you can polish the lenses to restore clarity, and applying a protectant can prevent the lenses from dulling too quickly.

WEATHER STRIPPING AND SEALS

The rubber weather stripping around the doors and windows creates a seal to keep out moisture, noise, and debris. Over time, the rubber may begin to deteriorate due to low humidity, contact with solvents, and air pollution. The seals can also freeze to the door frame in cold weather and tear when the door is opened.

To keep your weather stripping in good shape, take care not to tear it when entering or exiting the vehicle, and be cautious when opening doors in cold weather. If a seal is frozen, pour hot water on it to melt the ice. In dry climates, you may want to use a rubber conditioner to keep the weather stripping soft and elastic.

WINDSHIELD WIPERS

The wiper consists of the rubber blade, the frame that holds it to the windshield, the arm that swings it back and forth, and the motor that moves the arm.

You know you have a problem with your wipers if they aren't doing their job. Most problems, such as streaking or missed spots, are due to dirty, worn-out blades, or bent frames that aren't holding the blade against the windshield properly. The best way to extend the life of your wipers is to inspect and clean them each time you put gas in the car.

If the wipers aren't moving, check for a broken fuse. If you can hear the motor operating but the arms aren't moving, something has broken in the linkage to the wiper arm, and will require repairs.

Washing and Waxing Your Vehicle

Just as your paint protects the metal surfaces of your car, washing and waxing protects the paint. Pollution, acid rain, dirt, salt, tree sap, and many other things can deteriorate the paint. Keeping your car clean will help prevent premature corrosion and paint failure.

SUPPLIES

There are many different things you can use to clean your car. These are some of the basics. The most important thing to keep in mind is that all supplies should be clean. If you use a gritty sponge or dirty bucket, that debris will be transferred to your car and could damage the finish.

Soap Use a soap designed for washing vehicles. These soaps are mild and wash away cleanly. Avoid using household dish soaps and other liquid soaps, as they are designed to strip grease and oil and may be too harsh for the finish of your car.

Bucket You'll need a large bucket for soapy water. Make sure the bucket is clean and free from debris. You might consider designating a specific bucket just for car washing.

Sponge or Towels Wash your car with a soft sponge or towel. Terrycloth-covered sponges are a great option.

Soft-Bristle Brush A soft-bristle brush can be useful for cleaning convertible tops and for lifting stubborn debris from wheels.

AUTOMATIC CAR WASHES

If you choose to use an automatic car wash, make sure the equipment is clean. Dirty or ragged brushing surfaces can damage the exterior of your car. Touchless car washes use high-power sprays to blast the dirt off, but may not get into all the areas.

Polishing Cloths When applying wax or polish, use a polishing cloth. These cloths very soft, so they won't scratch the finish. If you select a microfiber cloth, make sure it is specifically designed for polishing.

Polish and Glaze Polishes are mildly abrasive and they take off minute portions of paint when applied. Polishing should be done sparingly. Glaze is an oily product that is non-abrasive. It fills in small swirl marks and scratches, making a smooth surface for applying wax. Polish and glaze are both optional steps when cleaning and waxing your car.

Bug and Tar Remover There are several products available that are designed to remove sticky road tar and bugs. They are usually an emulsion that penetrates and breaks up the bug matter or tar so it can be safely wiped away. Follow the directions; these products may be harmful to paint if left on too long.

Wax There are many different types of wax, including liquid, paste, carnauba, and silicone. Any of these can be used to protect the paint from the elements and environment. The thin layer of wax not only makes the paint shine, but also prevents oxidation and repels water and pollutants.

How to: Wash a Car

Washing your car regularly will not only keep it looking nice, it will also extend the life of the exterior.

WHAT YOU NEED

- Bug and tar remover
- Soft cloths
- Garden hose
- Soapy water
- Sponge or towels

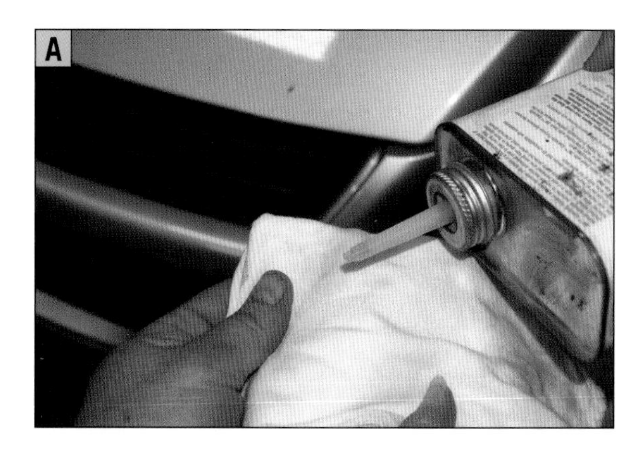
A

B

C

1. PARK IN A SHADY SPOT.
   Washing and waxing should be done out of direct sunlight, when the body of the vehicle is relatively cool. If water dries too fast, it can leave water spots that are difficult to get off.

2. REMOVE STUCK-ON RESIDUE.
   Apply a commercially available bug and tar remover to a soft applicator, like a polishing cloth (FIGURE A). Dab the stuck-on particle, let the remover work in a bit, and then gently rub the particles away. Don't allow the remover to sit too long on the paint.

3. RINSE WITH CLEAN WATER.
   Use a garden hose to rinse the entire vehicle.

4. WASH FROM THE TOP DOWN.
   Wash the car using a soft towel, wash mitt, or sponge and a bucket of soapy water. Start from the top and work your way to the bottom. Rinse frequently to get rid of the dirt and contaminants on your towel or sponge so they don't scratch the finish as you wash (FIGURE B).

5. RINSE AND DRY.
   Rinse the car completely with clean water. Then dry from the top down using a soft cloth or water-absorbent towel. If you run out of time at the bottom, you can lightly re-wet the spotted area without affecting the top of the car. Your towel will collect dirt and grease as you dry, so keep it rinsed (FIGURE C).

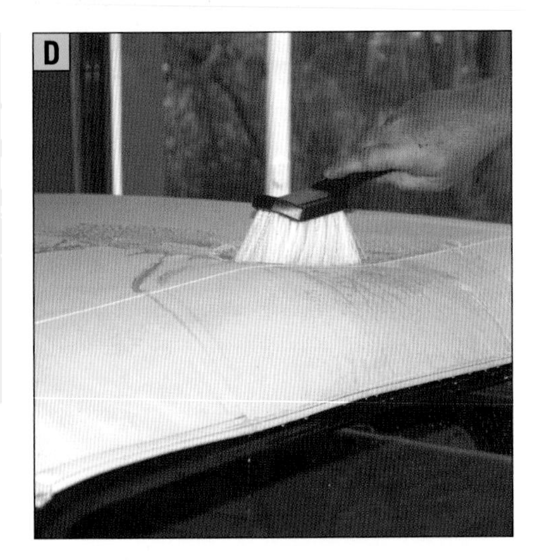
D

E

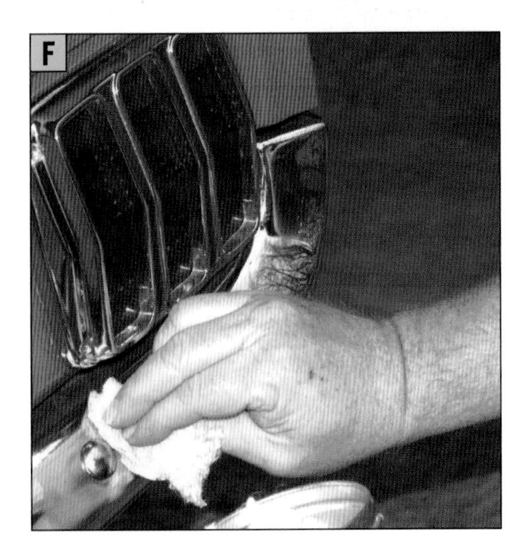
F

CLEANING VINYL TOPS

Use a soft-bristle brush and soapy water to clean vinyl or fabric convertible tops (FIGURE D). Check for mold in the creases and use a product designed for mold removal if necessary. Before using any soap or cleaner on a convertible top, test it on a small, inconspicuous spot.

CLEANING WHEELS

In addition to dirt from the road, wheels can get battered by the brake dust that comes off of disc brakes. This dust is very hot and can embed itself in the wheels.

When you clean your wheels, start with a gentle cleaning method before moving to more aggressive ones. Use soapy water and a soft cloth to scrub the wheels and tires from the outside into the center (FIGURE E). Clean the insides of the wheel well at this time, too. From there, use a soft-bristle brush to try to lift the particles out of the wheel. On chrome wheels, you can use super-fine steel wool, chrome polish, or chrome wadding. If you still have stubborn spots, use a wheel cleaner specifically made to remove brake dust.

Use caution when cleaning wheels; most aluminum and chrome wheels have a clear coating, and anything you apply to remove the embedded brake dust will probably damage the clear coat in the process. Never apply cleaners or detergents to a hot wheel surface; they can stain the wheel or remove the clear coatings.

After removing the dust and grime, coat the wheel with wax to help prevent brake dust from sticking to the wheel.

CLEANING CHROME

The chrome plating on your car needs attention or it can begin to pit and rust. Since chrome is a hard material, you can use a super-fine steel wool to remove dirt and rust and shine it up (FIGURE F). Chrome polish or wax should be applied after cleaning to keep out moisture and limit rust.

How to: Wax a Car

Wax your car in the shade and with the body panels cool. Whether you decide to polish your paint or glaze it before waxing is up to you, but make sure your car is thoroughly washed and clean before beginning. Work in sections using the following steps.

WHAT YOU NEED

- Polishing cloths
- Wax

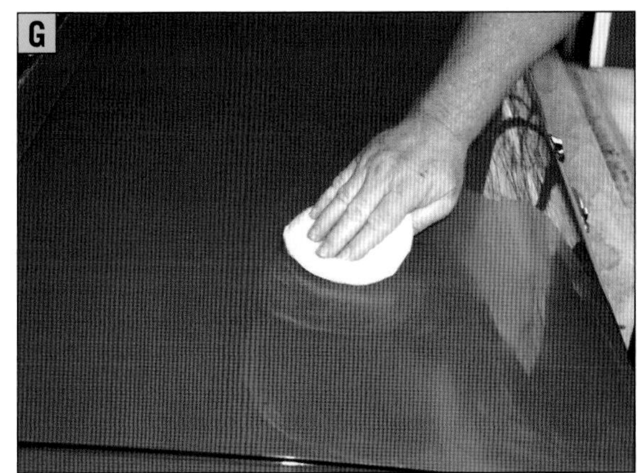

G

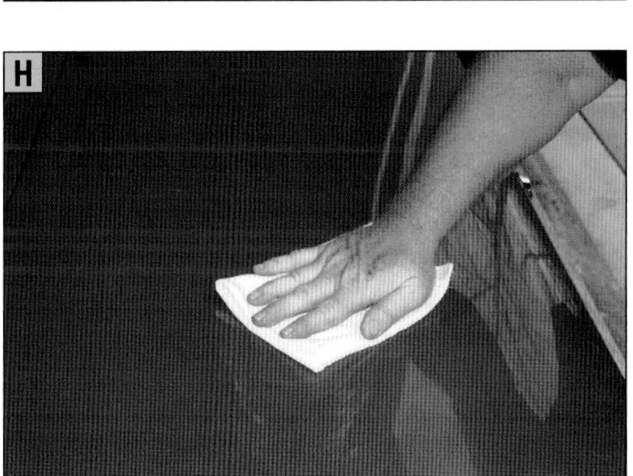

H

1. APPLY THE WAX.
   With an applicator pad or a polishing cloth, apply a thin layer of wax to a small area. Apply wax in a circular motion, using light but firm pressure (FIGURE G). Allow the wax to dry to a hazy film.

2. REMOVE THE WAX.
   Use a clean, dry polishing cloth to remove the film of wax. Apply light but firm pressure, and use a circular motion (FIGURE H).

3. BUFF TO A SHINE.
   After removing the wax, use a clean section of cloth to buff the finish to a shine, using the same pressure and a circular motion to get the best results.

4. REPEAT ON REMAINING SECTIONS.
   Work your way around the car, waxing and buffing one small section at a time.

POWER BUFFERS

Orbital buffers are great for getting a nice, shiny finish on your car when polishing or waxing, and they save your arms from all that manual work. If you choose to use one, take care—it is very easy to buff the paint off of edges, and if it gets clogged, you can heat the paint and burn through easily. Make sure the "bonnets" you use are designed for polishing or waxing paint.

Paint Imperfections and Rust Problems

Over time, your car's exterior will begin to show wear, either from minor impacts or simply from exposure to the elements. After a few years, you may begin to notice small paint imperfections or rust spots. Here's how to deal with them.

PAINT IMPERFECTIONS

The finish on your car consists of the paint, which provides the color, as well as a clear coat, which has protective properties. The clear coat finish is very thin—about the thickness of six human hairs—but it plays an important role in protecting your vehicle. If the clear coat or paint is compromised, rusting and further damage can occur.

Chips When the paint chips off and the underlying material of the car is exposed, the only solution is to replace or repaint the damaged part. If you notice a chip, take the car to a body shop for repair sooner rather than later.

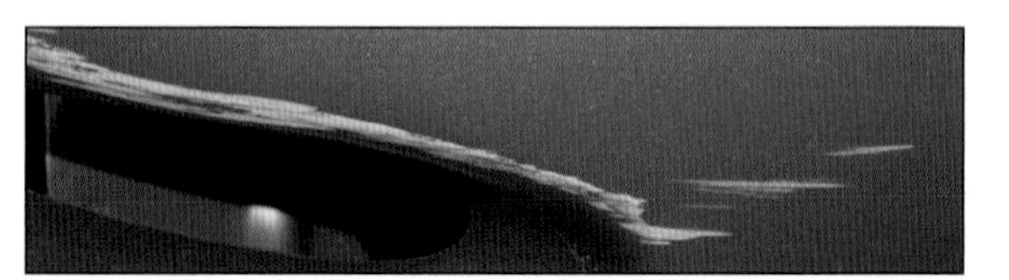

Scratches Even the most conscientious drivers occasionally make unintended contact with a mailbox, trash can, or the frame of the garage door. These minor impacts often result in white scuffs or scratches in the clear top coat. You may be able to remove shallow scratches like these on your own.

REPAIRING PAINT DAMAGE

Minor scuffs in your paint can be minimized or repaired by treating the surface with an abrasive product and filling in scratches with glaze. The most important rule in repairing paint is to begin with the least aggressive method possible and work your way up from there.

There are a wide variety of products that can be used to protect, maintain, and repair paint. The following are some of the most commonly used, from least to most abrasive.

Wax Wax is not abrasive. It is used to not only make the paint shiny, but also to protect it from oxidation, pollutants, and minor contact. Waxing every three months or so will help to extend the life of your car's finish.

Clay Clay bars are used to lift dirt and contaminants away from the paint surface before the car is polished. Clay does not take out swirl marks or scratches.

Glaze This term is used for many different products, but the characteristics most glazes share are that they are oily and non-abrasive. Glaze is designed to fill in small swirl marks and scratches before waxing. It creates a nice, even surface for wax application.

Polishes Polishes are mildly abrasive and are meant to take out very small scratches, swirl marks, and imperfections in the paint. Because of their abrasive quality, polishes do remove a very small amount of paint, so use them sparingly.

Compounds Compounds are more abrasive than polishes and are meant for taking out deeper scratches. Remember, the more aggressive the abrasive, the more paint is removed.

Polishes and compounds can be in liquid or paste form. The more aggressive the abrasive, the grittier the product will feel.

Sandpaper Sandpaper should be a last resort when repairing paint damage. Even when using a very fine grit (2000 or 3000), it is easy to sand too deeply and break through the clear topcoat. Sandpaper made for automotive paint is generally much finer than sandpapers for wood, and can be used wet or dry.

RUST

Rusting occurs when the metal parts of your car come into contact with moisture and oxygen. Rust spreads quickly, so take steps to stop it if you can.

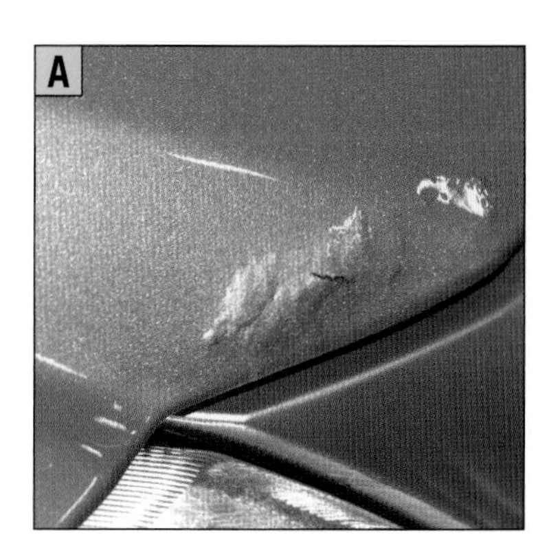

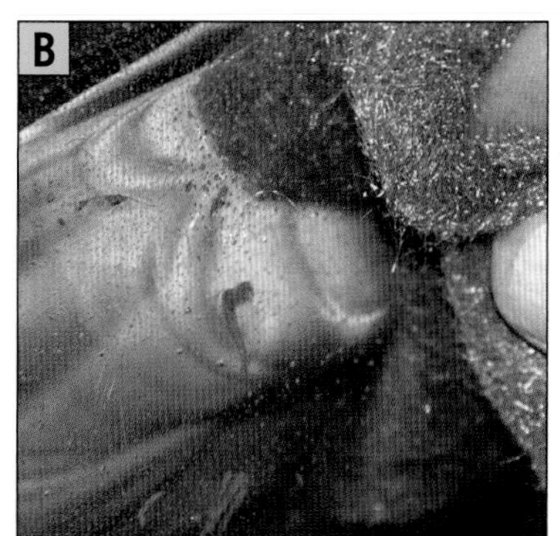

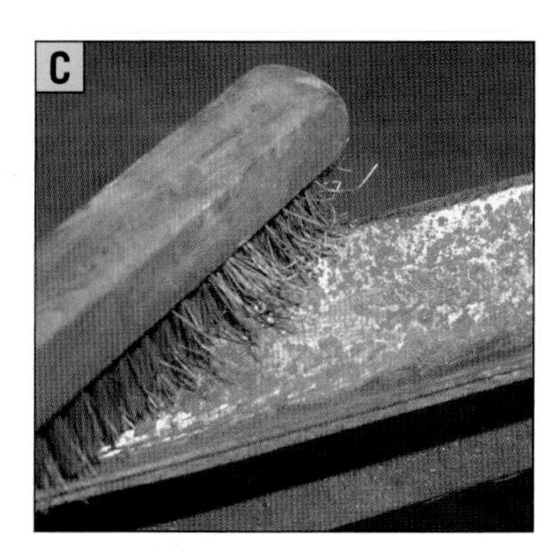

Rust Bubbles Rust bubbles appear when a break in the paint causes it to pull away from the body (FIGURE A). This kind of rust needs to be removed and repaired. If there is an opening in the bubbles, squirt a little motor oil into the area to help prevent the rust from spreading. The oil will seep down and displace any moisture, reducing the spread until you can get it fixed. Remember to reapply the oil after washing the car.

Bumper Rust Chrome parts rust easily, but they can withstand more abrasion than paint. Clean off the rust with super-fine steel wool, and then use a chrome polish to prevent further oxidation of the area (FIGURE B).

Frame Rust The frame of the car is sturdy, so you can scrape off the rusted area with a stiff metal brush and coat it with paint or a rust encapsulator to prevent further damage (FIGURE C).

How to: Remove Scuffs in Paint

Minor impacts can often leave white streaks in the clear coat. These can be removed or reduced using glaze, polish, and possibly sandpaper. The key is to use the least aggressive method first and move up from there.

WHAT YOU NEED

- Soapy water and towel
- Polishing cloths
- Glaze
- Polish

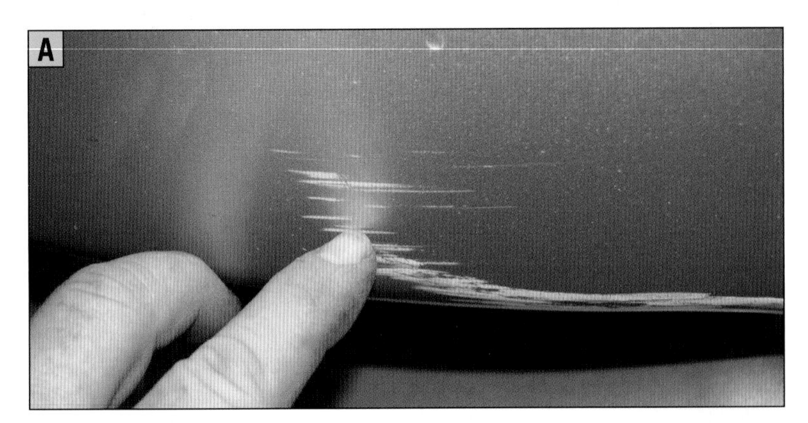

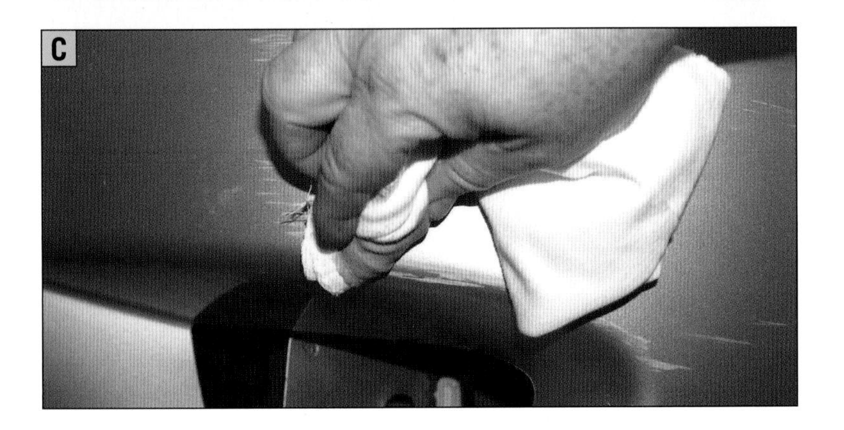

1. ASSESS THE DAMAGE.
   Run your fingernail across the scratches to see how deep they are. This scuff is fairly mild, but it does have some deeper scratches that catch on the nail (FIGURE A). Chipped paint or scratches that go through the clear coat layer will need to be repainted.

2. WASH THE AREA.
   Thoroughly wash and dry the area where you will be working so it is free of dirt and contaminants (FIGURE B).

3. APPLY GLAZE.
   Put a small amount of glaze on a soft polishing cloth and slowly rub it into the scuff (FIGURE C). Work in the direction of the scratches if you can.

   After glazing, only the biggest scratches should be visible.

D

E

4. POLISH THE BIGGER SCRATCHES.
   Apply mildly abrasive polish to larger scratches (FIGURE D). Follow the product instructions, but in general, use a circular motion to work it into the paint. You may need to apply more than one coat of polish.

5. APPLY A SECOND COAT OF GLAZE.
   Another coat of glaze will help to fill any light abrasions left by the polishing (FIGURE E).

TIP

Be patient; completing several applications using a gentle method is better than jumping to a more aggressive method too quickly. You'll be less likely to cause permanent damage if you take your time.

How to: Restore Headlight Lenses

Most vehicles have a clear, polycarbonate lens covering the headlights. Over time, oxidation, UV rays, and road debris can dull the lens and reduce the amount of light coming out of the headlights.

WHAT YOU NEED

- Fine-grit automotive sandpaper (1000, 2000, and 3000 grit)
- Plastic lens polish
- Power drill and buffing pad (optional)
- Polishing cloths
- Headlight protective coating

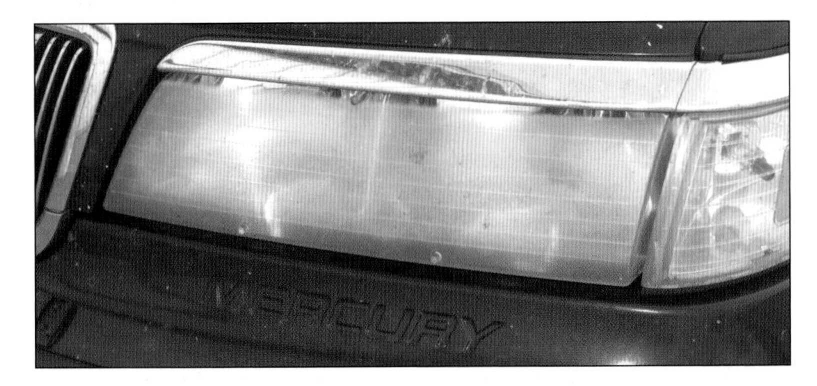

This car's headlight lenses have suffered from years of sunlight exposure, road grime, and oxidation. We'll sand and polish half of the lens to show the results.

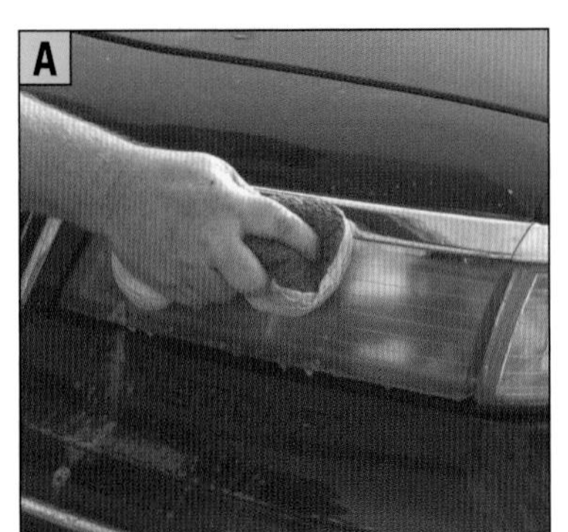

A

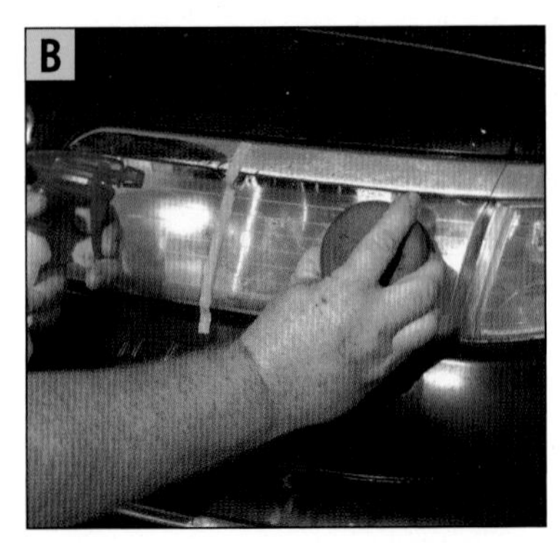

B

C

1. CLEAN THE HEADLIGHTS.
   Remove all the dirt and grime before starting, so you aren't rubbing the dirt into the lens (FIGURE A).

2. SAND THE LENS.
   Start with the 1000-grit sandpaper. Dampen the sandpaper and keep it wet to draw away the sanding particles. A spray bottle with water will help keep both the lens and the sandpaper rinsed (FIGURE B).

3. CONTINUE SANDING WITH FINER PAPER.
   After the 1000-grit paper, use 2000-grit and then 3000-grit until you have an even haze on the lens (FIGURE C).

CAUTION

Be careful when using power drills and pressure; the lens can heat up, causing the plastic to warp or burn. Also take care to avoid scratching the paint with sandpaper and polish.

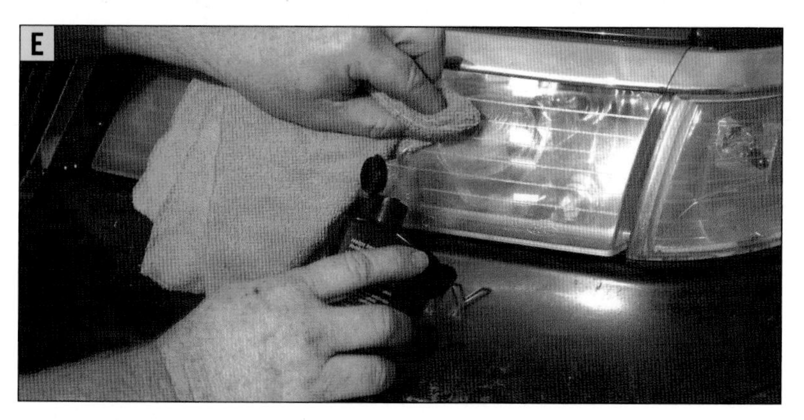

4. **POLISH THE LENS.**
   Wipe the lens with a damp cloth and then apply plastic polish either by hand with a polishing cloth, or with a power drill and buffing pad. Use light but firm pressure when using a drill, and keep the buffer moving evenly (FIGURE D).

5. **APPLY PROTECTANT.**
   Apply lens protectant with a polishing cloth. Use it regularly to keep your headlights from dulling too soon (FIGURE E).

   After sanding and polishing, the lens is much clearer and the headlight is brighter (FIGURE F).

How to: Inspect and Replace Wiper Blades

Windshield wiper blades should be replaced when they stop functioning properly, or every 6 to 12 months. The thin, flexible rubber of the blade breaks down over time due to exposure to the elements, as well as normal use.

WHAT YOU NEED

- Paper towel
- Replacement wiper blades

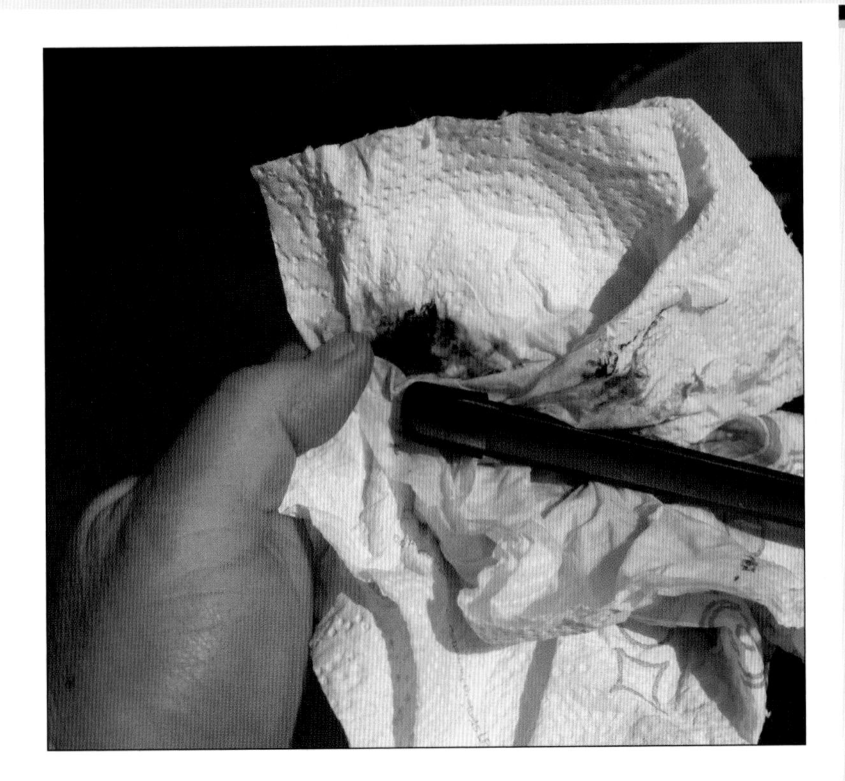

INSPECTING WIPER BLADES

Each time you fill up your car, use a paper towel to wipe the blade clean of oils, dirt, loose rubber, and anything else that could smear on your windshield. Flex the blade to make sure it is soft and pliant, and check for cracked or broken rubber. Check the metal frame for rust or broken parts.

TIPS FOR MAINTAINING WINDSHIELD WIPERS

- If possible, pull your wipers away from the windshield when your car is parked outside in icy or snowy weather. This prevents them from freezing to the windshield, and makes it easier to scrape your windshield without gouging the rubber of the blade.
- Don't use your wiper blades to clear ice or snow. Running them up and over the ice can tear the thin rubber. Use heat from the defroster to loosen frozen wiper blades.
- Always use a washer fluid that contains antifreeze. It helps to clean debris, de-ice the windshield, and is better at keeping your blades clean.

REPLACING WIPER BLADES

Make sure to purchase the correct replacement blades for your vehicle. Auto parts stores that sell replacement blades will have a chart or book where you can look up the type and length of blade you need.

1. **REPOSITION THE WIPERS IF NEEDED.**
   You may need to turn on the car to adjust the position of the wipers (FIGURE A). Make sure you can lift them away from the windshield easily. Turn the car off before changing the blades.

2. **RELEASE THE WIPER BLADE.**
   Most wipers use one of two methods to release the old blade. One is spring loaded, and the other employs a small tab on the underside of the wiper arm that you push down to release the blade (FIGURE B). Your new wiper blades should have instructions on which ones to use on your car.

3. **REMOVE THE OLD BLADE.**
   Pull the wiper arm away from the windshield to get the proper angle for removing the old blade. With the tab or spring in the release position, slide the blade back to release it from the bracket (FIGURE C). You may have to work a little bit to relieve the arm of the blade assembly.

4. **INSTALL YOUR NEW BLADE.**
   Installing the new blade is the opposite of removal. Pull the arm away to attain the correct angle for installing the blade, and slide it on until you hear it click into place (FIGURE D). Slowly move the arm back into position on the windshield and check to ensure that the frame is providing even pressure on the blade and windshield.

How to: Inspect and Repair Weather Stripping

The rubber weather stripping around car windows and doors keeps noise, dirt, and moisture out of your car. Over time, it can dry out or deteriorate, making it less effective. Inspect weather stripping regularly and replace it if it is cracked or torn. Weather stripping is usually held in place with plastic pins or glue.

WHAT YOU NEED

- Pliers or screwdriver
- Replacement pins (if used)
- Weather stripping adhesive (if used)
- Rubbing alcohol (if needed)
- New weather stripping (if needed)
- Gloves

INSPECTING WEATHER STRIPPING

You should inspect the weather stripping on your vehicle once or twice each year. Check around the doors, trunk lid, and windows. Be sure to check the underside of the doors; weather stripping in this area is usually hidden from view and frequently comes loose.

Inspect the weather stripping for damage, and to see if the rubber is becoming hard and brittle. Give it a gentle tug to make sure it is securely fastened, and check to see if the seal has been flattened and needs replacing.

REPAIRING WEATHER STRIPPING AFFIXED WITH PLASTIC PINS

Some manufacturers use plastic, one-time-use pins to fasten the weather stripping in place. These pins can break or be torn out, leaving the weather stripping unattached.

Before beginning a repair with plastic pins, remove one of the old pins and take it to an auto parts store to ensure that you buy the correct replacement pins. Many pins look alike, so it's best to bring an example.

1. REMOVE THE OLD PINS.
   Hold the pin in place with a pair of long-nose pliers, and gently stretch the weather stripping away from the pin. It should slide easily out of the hole.

2. INSTALL THE NEW PINS.
   Put pressure behind the pin base and press it into the hole until the tangs catch.

3. DOUBLE CHECK THE FIT.
   Check that the seal is secure against the original mount, and make sure you didn't pull out any additional pins while installing the new ones. Close the door or window to make sure there are no obstructions.

REPAIRING WEATHER STRIPPING AFFIXED WITH ADHESIVE

For this repair, you'll want to use an adhesive that is specifically made for gluing or repairing rubber weather stripping. Household glues won't hold up to the weather, and some may actually dissolve the rubber. You can find weather stripping adhesive at most auto parts suppliers.

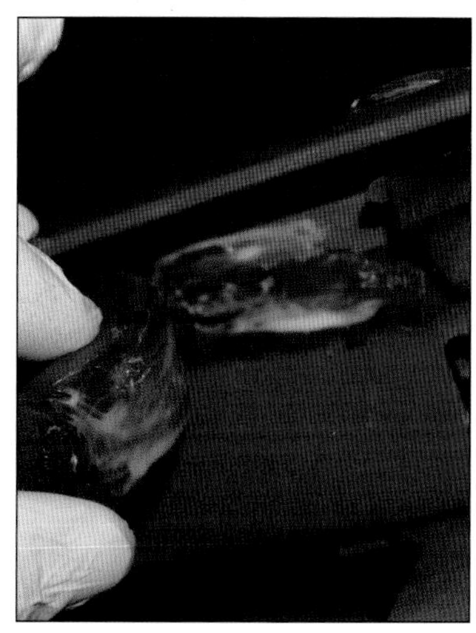

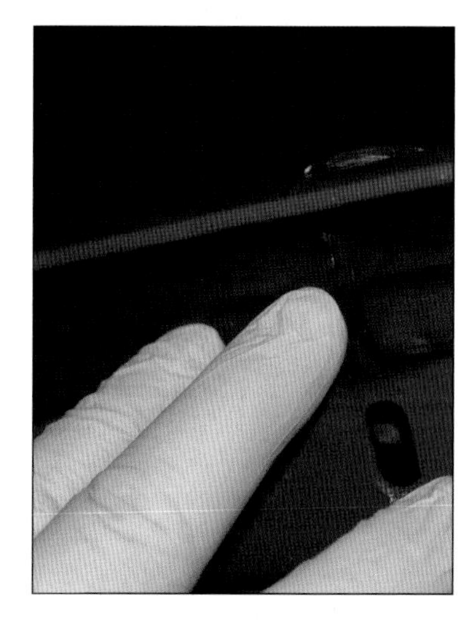

1. PREPARE THE AREA.
   If you are reattaching old weather stripping, remove any excess glue and debris using rubbing alcohol. Stubborn glue may require adhesive remover (FIGURE A). You can also clean new weather stripping with rubbing alcohol to remove any residue left over from the manufacturing process.

2. APPLY A THIN BASE LAYER OF ADHESIVE.
   Using a gloved finger or a small brush, apply a thin layer of adhesive to both the weather stripping and the surface to which it will be glued and let the adhesive dry (FIGURE B). This helps the adhesive bond to the parts.

3. GLUE AND PRESS INTO PLACE.
   Apply another thin layer of adhesive to the weather stripping and press it firmly into place (FIGURE C). Hold it for a minute or two, and then close the window or door. The pressure of the closed seal will hold the piece in place until the adhesive has dried completely.

TIP

You can use weather stripping adhesive to make minor repairs to torn weather stripping, but the repair may not last as long as installing a new seal.

Maintenance Schedules

This is a general guideline for routine maintenance tasks. Check your owner's manual for any recommended schedules that are specific to your vehicle.

EACH TIME YOU DRIVE

- ☐ Inspect tires for problems.
- ☐ Check for exterior damage.
- ☐ Look for leaks.

EACH TIME YOU STOP FOR FUEL

- ☐ Check oil.
- ☐ Check tire air pressure.
- ☐ Clean wipers and windshield.
- ☐ Inspect belts and hoses.
- ☐ Check fluid levels.

EVERY 3,000 MILES (5,000 KM) OR THREE MONTHS

- ☐ Complete the fill-up checks.
- ☐ Change oil.
- ☐ Inspect tires for wear.
- ☐ Check fluid levels and condition.
- ☐ Lubricate chassis.
- ☐ Inspect belts and hoses.
- ☐ Check for engine leaks.
- ☐ Check air filter.
- ☐ Check external lights.
- ☐ Check battery terminals.
- ☐ Run fuel injector cleaner.
- ☐ Wax vehicle.

EVERY 6,000 MILES (10,000 KM) OR SIX MONTHS

- ☐ Complete the 3,000 mile checks.
- ☐ Rotate the tires.
- ☐ Check automatic transmission fluid.
- ☐ Check seat belts.
- ☐ Check fuel cap.
- ☐ Check seals.
- ☐ Check computer codes.
- ☐ Check wheel alignment.

EVERY 12,000 MILES (20,000 KM) OR 12 MONTHS

- [ ] Complete the 6,000 mile checks.
- [ ] Replace air filter.
- [ ] Inspect PCV valve.
- [ ] Inspect fuel system.
- [ ] Inspect spare tire.
- [ ] Check cabin filter.
- [ ] Check battery.
- [ ] Inspect brakes.

EVERY 24,000 MILES (40,000 KM) OR TWO YEARS

- [ ] Complete the 12,000 mile checks.
- [ ] Replace fuel filter.
- [ ] Replace transmission fluids and filter.
- [ ] Replace the coolant.
- [ ] Inspect PCV valve and replace if needed.
- [ ] Inspect brakes and linings.

EVERY 48,000 MILES (80,000 KM) OR FOUR YEARS

- [ ] Replace timing belt.
- [ ] Inspect spark plugs and replace if needed.
- [ ] Replace engine belts.
- [ ] Check differential fluids.
- [ ] Check shocks/struts.
- [ ] Check battery output/load.

EVERY 60,000 MILES (100,000 KM) OR FIVE YEARS

- [ ] Replace spark plug wires.
- [ ] Replace shocks.
- [ ] Flush coolant and replace.
- [ ] Replace battery (conventional engine).
- [ ] Check batteries (hybrid).

Terms to Know

anti-lock brakes (ABS) A safety feature that uses a sensor to detect if the wheel has stopped moving prematurely. The computer will release pressure to the wheel and "pulse" it on and off to slow the car safely and quickly.

aftermarket Term used to describe item that didn't come from the original manufacturer of your car, such as tires, fluids, performance parts, fancy wheels, etc.

amps Short for *ampere*, which is the flow of an electric charge. It is the flow of electrons through a circuit, pushed along by voltage.

catalytic converter A device in the exhaust system that converts toxic emissions into less toxic emissions.

continuity An electrical connection that is complete and passing electrical power.

combustion The explosion that occurs in the motor when air and fuel are mixed together, compressed, and ignited.

compression How much your engine can squeeze the air and fuel in the cylinder. The more compression, the more power can be gleaned from the combustion.

coil-on-plug (COP) An ignition system format in which each spark plug has its own ignition coil.

CV joint An abbreviation for *constant-velocity joint*. This type of joint allows elements that are not aligned to turn together. CV joints are protected by a rubber boot.

directional tread A tire that is designed to be run on one side of the car. It cannot be rotated to the other side.

DLC Abbreviation for *data link connector*. This is the D-shaped connector used to talk to the computer in your car.

drivetrain How the engine gets the power to the wheels. This usually refers to the transmission, the drive wheels, and axles.

dry rot A form of deterioration that affects rubber components in a vehicle.

E-85 A gasoline that is a blend of 85 percent ethanol or bio-fuel, and 15 percent normal gasoline.

EGR Abbreviation for exhaust gas recirculator. The engine takes a portion of the exhaust gases and sends it back into the engine to be burned again.

emissions The results of burning fuels. Different fuels will emit different types of emissions.

EVAP Abbreviation for *evaporative emissions control system*. The fuel vapors in the fuel tank are drawn into the engine and burned to reduce emissions.

HID bulb A type of headlamp bulb that charges a gas like xenon instead of using a filament.

horsepower The measure of how much work your engine can do.

hybrid A vehicle that uses both a fuel-burning engine and electrical power. The car is driven by the electric motor, and depending on the design, the engine can generate power for the electric motors, provide electricity to drive the wheels, or both.

kilopascal (kPa) The metric measure of pressure. It is used to measure pressure of air and fluids in your car.

LED bulb Abbreviation for *light emitting diode*. LED bulbs run cooler and longer lasting than traditional incandescent bulbs.

load The resistance on something trying to do work. An empty car sitting at idle has less load than a full car trying to go up a hill.

O_2 sensor A sensor that measures the emissions from the engine and the catalytic converter. The computer uses the information to adjust the amount of air, fuel, and timing.

OBDII Abbreviation for *on-board diagnostics II*. This is the set of error codes most cars use to tell you or the mechanic of problems with the car.

octane A measure of how much a fuel can be compressed before it self-ignites. The higher the number, the more it can be compressed and the more power can be achieved.

PCV valve Abbreviation for *positive crankcase ventilation valve*. A one-way valve the allows the pressure and vapors building up inside the engine to be released and sent into the engine to be burned.

powertrain The engine, transmission, drive axles, gears, and wheels are considered the powertrain.

pre-detonation When the fuel and air mixture starts to ignite before the engine is ready. The result is lost power and damage to engine parts.

psi Abbreviation for *pounds per square inch*. Used to measure the pressure of air and fluids in your car.

resistance How much something prevents electrical power from passing through it. Resistance is measured in ohms (Ω). The more ohms, the harder it is to pass electricity through the item.

Schrader valve The type of valve used to fill the A/C system and check the fuel pressure.

TPMS Abbreviation for *tire pressure monitoring system*. Small pressure monitors are mounted in the tires and send a signal to the computer if the pressure in the tires gets too low.

torque Ability to turn something, like the wheels or the engine. The more torque, the easier it is to turn.

torque spec or specification How much force it takes to stretch a thread on a bolt, nut, or fastener so it holds without backing off. Too much torque can break a fastener, and not enough may allow it to come loose. Torque spec is dependent on the size, material, coating, and hardness of the fastener.

U-joint Abbreviation for *universal joint*. An X-shaped part with rotating bearings on the end to allow two rotating items to work together when they are not straight to each other. Similar to a CV joint.

viscosity The thickness of a fluid. Viscosity is measured with a "w" (weight) number. The higher the weight, the thicker it is.

volts or voltage The measure of electrical pressure difference in an electrical circuit. Think of voltage as a big pump that pushes the electrical current (amps) along.

Zerk fitting A small one-way valve used on components of your car to grease joints without letting the grease come back out.

Index

C

D

J–K

L

M

N

O

T

U–V

W–X–Y–Z

All photos by Dave Stribling, with exception of the following:

p. 49: Ben Stoner, Fathouse Fabrications

pp. 171, 183: William Thomas

p. 196: Ann Barton

p. 201: Mike Dunning © Dorling Kindersley, Courtesy of The Science Museum, London